Mathematics and Numbers of the Prophecy

The Second Trumpet

Andrew the Prophet

Online lessons available at andrewtheprophet.com

iUniverse, Inc.
New York Bloomington

Mathematics and Numbers of the Prophecy
The Second Trumpet

iUniverse books may be ordered through booksellers or by contacting:
iUniverse
1663 Liberty Drive
Bloomington, IN 47403
www.iuniverse.com
1-800-Authors (1-800-288-4677)

Printed in the United States of America

ISBN: 978-1-4401-1984-2 (pbk)
ISBN: 978-1-4401-1983-5 (ebk)

Library of Congress Control Number: 2009901505

iUniverse rev. date: 3/9/2009

Table of Contents

Axiom on Truth and Knowledge

To Andrew the Prophet

Completed March 4, 2008

"I do not know what I may appear to the world, but to myself I seem to have been only a boy playing on the sea shore, and diverting myself now and then finding a smoother pebble or a prettier sea shell than ordinary whilst the great ocean of truth lay all undiscovered before me."

Sir Isaac Newton

Truth is more apparent than men's thoughts may reveal. For God created man with the freedom of thought, that all men may search for the truth. For God created this world in the image of the truth. And the truth is discovered through knowledge. And knowledge can be found through mathematics and numbers.

Mathematics is the science of logical patterns, and truth is established by deduction from these patterns. Yet most assume that mathematics is a science, yet science is limited by the laws of this universe. And though mathematics pertains to these laws, it extends beyond the laws of the universe. For the beauty of mathematics lies not in its complexity, but in its simplicity and the elegance of its truths. And the most elegant of math's truths exists in its foundation, and its foundation is based upon numbers.

Numbers are symbols which are utilized for measuring and counting. And as Plato expounded, "numbers are the highest degree of knowledge, for it is knowledge itself." And the simplest of numbers is natural numbers founded in Mesopotamia in 3400 BC. And through the Mayan and Hindu civilizations, the abstraction of numbers was founded through the conception of zero. But the most beautiful and complex of numbers are the transcendental numbers of e, i, and π. For the mystery of numbers and the universe can be discovered in Euler's formula:

$$e^{i\pi} + 1 = 0.$$

And though knowledge is the beginning of truth, and truth is the knowledge of things as they are, man has blinded himself with ignorance. For "ignorance is the root and stem of every evil". (Socrates) "Is it not a bad thing to be deceived about the truth, and a good thing to know what the truth is? For I assume that by knowing the truth you mean knowing things as they really are." (Plato) For "three things are necessary for the salvation of man: to know what he ought to believe; to know what he ought to desire; and to know what he ought to do." (St. Thomas Aquinas) But man has despised knowledge and truth, and in ignorance has continued to sin against God. For through blinding ignorance man does not see, for "if you were blind, you would have no sin; but since you say, 'We see,' your sin remains." (John 9:41) So let us strive to find knowledge and truth, for truth can be found in the foundation of "man's truth" – for the truth exists in ***M***athematics ***A***nd ***N***umbers.

ατπ

Zero is the Circle Back to the Father (NASB)

To Andrew the Prophet

Completed January 5, 2008

"When He established the heavens, I was there, when He inscribed a ***circle*** on the face of the deep, when He made firm the skies above, when the springs of the deep became fixed , when He set for the sea its boundary so that the water would not transgress His command, when He marked out the foundations of the earth." (Proverbs 8:27-29)

The Bible does not contain the number zero, nor does the Quran, nor the Greek alphabet, nor the Roman numerical system, nor the Egyptian hieroglyphs. For the concept of zero originates from the mathematics of India. But unlike the agnostic notions of zero, their concept of zero was fundamentally correct. For zero was founded upon the belief, that the earth was created in the image of the heavens, and that man's goal was to renounce the materialism of this world, and to return to the realm of eternity. And as the Son did say "whoever wishes to save his life will lose it, but whoever loses his life for My sake and the gospel's will save it. For what does it profit a man to gain the whole world, and forfeit his soul?" (Mark 8:35-37) Thus the circle of returning to the realm of the Father, was the original and correct concept of zero.

Zero was introduced to the west through India, by the Italian mathematician Fibonacci. The original word zero comes from the Arabic word *sifr* which was translated into the Italian word *zefiro*, and was later translated into the Venetian word *zero*. For *zefiro* is derived from the word *zephyrus* which means "westerly wind." And the prophet foretold that the name of the Lord would arise from the westerly wind. "So they will fear the name of the Lord from the ***west*** and His glory from the rising of the sun, for He will come like a rushing stream which the ***wind*** of the Lord drives." (Isaiah 59:19)

And the number of zero has great symbolic significance. For the circle of zero has been used throughout history, to symbolize the realm of the heavens. For in Asia, the circle symbolizes Yin and Yang: the polarity of man and nature, of life and death, and their interdependence on one other. And in Nepal, the circle of the mandala is symbolic of the universe. And a true follower of Christ, lives in the hope of returning to the Father. For all men partake of the circle of life, with the hope of returning to the Creator. For the foundation of the heavens was established from the beginning. For as the Son affirmed, "when He established the heavens, I was there, when He inscribed a ***circle*** on the face of the deep, when He made firm the skies above, when the springs of the deep became fixed, when He set for the sea its boundary so that the water would not transgress His command, when He marked out the foundations of the earth." (Proverbs 8:27-29)

The circle of zero is formed from a polygon, and the foundation of all polygons is the three-sided triangle. And in the beginning, the Trinity formed the heavens and the earth. "In the beginning God (***the Father***) created the heavens and the earth. The earth was formless and void, and darkness was over the surface of the deep, and ***the Spirit of God*** was moving over the surface of the waters. Then God said, 'Let there be light'; and there was light (***the Son***)." (Genesis 1:1-3) And God has multiplied His creation in manifold ways, to complete the circle of His hands. And by multiplying the sides of a polygon by an infinite number, the polygon of the circle is completed. And when all things are completed on heaven and earth, the circle of God's creation shall be complete. "It is done. I am the Alpha and the Omega, the beginning and the end. I will give to the one who thirsts from the spring of the water of life without cost. He who overcomes will inherit these things, and I will be his God and he will be My son." (Revelation 21:6-7)

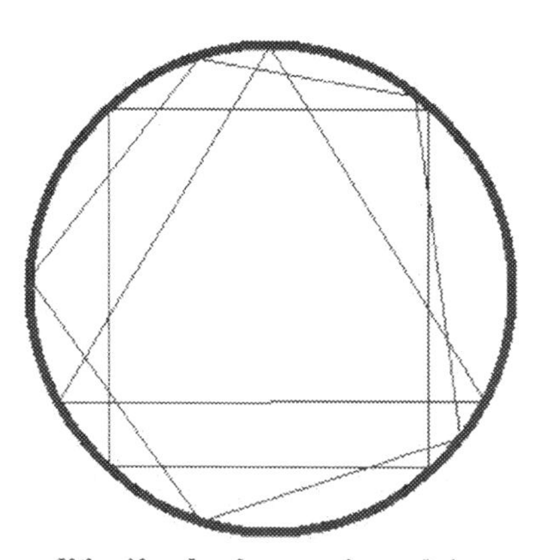

If the sides of a polygon equals n, and when n approaches infinity, then the polygon becomes a circle.

Madonna and Child

Christ and cruciform nimbus
Byzantine mosaic, 12th century

***Last Judgment,* Giotto 1305**

And the zero symbolizes the holiness of God: the halo over Christ, the "cruciform nimbus" over the Son, and the rainbow that surrounds the throne of God. "And He who was sitting was like a jasper stone and a sardius in appearance; and there was a rainbow around the throne, like an emerald in appearance." (Revelation 4:3)

And although the zero was established in India, a separate ancient civilization received the same revelation. For the Mayans referred to zero as a circle surrounding two gates, which symbolized the "spiral of life". And throughout the history of mankind, the spiral has been of great symbolic significance. For in China the spiral has symbolized the sun. And in many civilizations, the spiral was inscribed on their graves, to symbolize the circle of "life, death, and rebirth". And the spiral has been repeated throughout nature: on the spiral of shells, the spiral of DNA, the spiral of our fingerprints, and the spiral of the galaxies. For a spiral is defined as a geometrical entity, that rotates about the original point. And all creation rotates about one original point... and that point is the Creator who is GOD. "In the beginning God created the heavens and the earth." (Genesis 1:1)

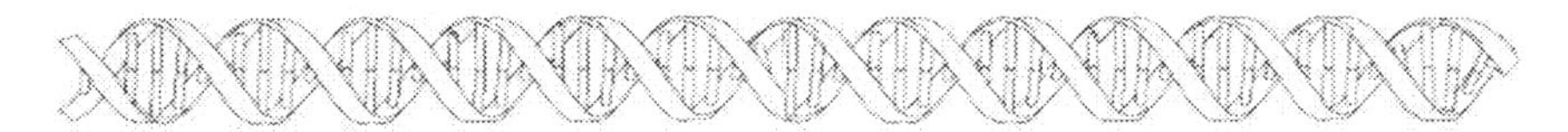

And a perfect spiral is known as a Fibonacci spiral. And this spiral is formulated from the Fibonacci number, which is a sequence which follows the mathematical progression:

$F(0) = 0$

$F(1) = 1$

$F(n + 2) = F(n) + F(n + 1)$ for all $n \geq= 0$

Thus the Fibonacci series progresses as follows: 0, 1, 1, 2, 3, 5, 8, 13, 21, 34, 55, 89, 144 ... And if we tile the Fibonacci series into squares, then we create the Fibonacci spiral. And the beauty of the Fibonacci spiral which replicates the Golden Ratio is seen throughout nature: as in the spiral of a shell and in the spiral of the universe.

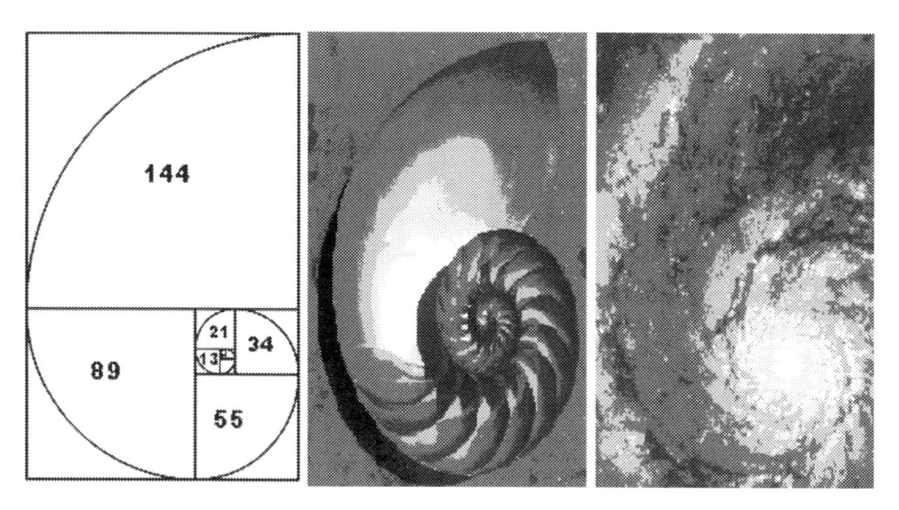

And the Mayan symbol for zero is a circle with two gates ($\pi\pi$). And the circle back to the Father is a circle with two Gates: the Lamb and the Lion. For **Christ the Lamb** is π, for He is the narrow gate back to the Father. (*Addendum 1*: *Christ is the Gate*) "For the gate is small and the way is narrow that leads to life, and there are few who find it." (Matthew 7:14) And **Christ the Lion** is π, for He is the wide gate back to the Father. "For the gate is wide and the way is broad that leads to destruction, and there are many who enter through it." (Matthew 7:13) For only through the Son is the circle made complete. For as He said "I am the way, and the truth, and the life; no one comes to the Father but through Me." (John 14:6) For the Son came as the Lamb but shall return as the Lion. And the gates of the Son shall make the circle complete. For "a ***lion*** has gone up from his thicket, and a destroyer of nations has set out; He has gone out from his place to make your land a waste. Your cities will be ruins without inhabitant. For this, put on sackcloth, lament and wail; for the fierce anger of the Lord has not turned back from us." (Jeremiah 4:7-8) "And the ***Lamb*** will overcome them, because He is Lord of lords and King of kings, and those who are with Him are the called and chosen and faithful." (Revelation 17:14) But woe to those who must enter through the Lion!

Mayan Symbol of Zero

The Gates of the Lion and the Lamb

But man is a fool because man has forgotten the meaning of zero. For rather than acknowledging that all things come from God, man in ignorance believes that he has control. For in numerology the zero is *The Fool*. And man in his folly has become the fool, for "the fool has said in his heart, 'There is no God.'" (Psalms 14:1)

Nothing is nothingness

ατπ

I

God is One (NASB)

To Andrew the Prophet

Completed November 21, 2007

God is One, there is no one else

“So that all the peoples of the earth may know that the Lord is God; there is no **one** else.” 1Kings 8:60

“Nothing on earth is like him, **One** made without fear.” Job 41:33

“The Mighty **One**, God, the Lord, has spoken, and summoned the earth from the rising of the sun to its setting.” Psalms 50:1

“The **One** forming light and creating darkness, causing well-being and creating calamity; I am the Lord who does all these.” Isaiah 45:7

“For your husband is your Maker, whose name is the Lord of hosts; and your Redeemer is the Holy **One** of Israel, who is called the God of all the earth.” Isaiah 54:5

“But God is the Judge; He puts down **one** and exalts another.” Psalms 75:7

“‘To whom then will you liken Me that I would be his equal?’ says the Holy **One**.” Isaiah 40:25

The Son of Man is the One

“I kept looking in the night visions, and behold, with the clouds of heaven **One** like a Son of Man was coming, and He came up to the Ancient of Days and was presented before Him. And to Him was given dominion, glory and a kingdom, that all the peoples, nations and men of every language might serve Him. His dominion is an everlasting dominion which will not pass away; and His kingdom is **one** which will not be destroyed.” Daniel 7:13-14

"And He said, 'The **one** who sows the good seed is the Son of Man'" Matthew 13:37

"But as for you, Bethlehem Ephrathah, too little to be among the clans of Judah, from you **One** will go forth for Me to be ruler in Israel. His goings forth are from long ago, from the days of eternity." Micah 5:2

"All things have been handed over to Me by My Father; and no **one** knows the Son except the Father; nor does any**one** know the Father except the Son, and any**one** to whom the Son wills to reveal Him." Matthew 11:27

"I and the Father are **one**." John 10:30

"For God, who said, 'Light shall shine out of darkness,' is the **One** who has shone in our hearts to give the Light of the knowledge of the glory of God in the face of Christ." 2 Corinthians 4:6

"But do not be called Rabbi; for **One** is your Teacher, and you are all brothers. _Do not call any**one** on earth your father; for **One** is your Father, He who is in heaven. Do not be called leaders; for **One** is your Leader, that is, Christ." Matthew 23:8-10

God gives us one heart through the Holy Spirit

"And I will give them **one** heart, and put a new spirit within them. And I will take the heart of stone out of their flesh and give them a heart of flesh." Ezekiel 11:19

"But the **one** who joins himself to the Lord is **one** spirit with Him." 1 Corinthians 6:17

"There is **one** body and **one** Spirit, just as also you were called in **one** hope of your calling; **one** Lord, **one** faith, **one** baptism,_**one** God and Father of all who is over all and through all and in all." Ephesians 4:4-6

"The Spirit and the bride say, 'Come.' And let the **one** who hears say, 'Come.' And let the **one** who is thirsty come; let the **one** who wishes take the water of life without cost." Revelation 22:17

We are called to be one flock and body with God

"All that the Father gives Me will come to Me, and the **one** who comes to Me I will certainly not cast out." John 6:37

"I have other sheep, which are not of this fold; I must bring them also, and they will hear My voice; and they will become **one** flock with **one** shepherd." John 10:16

"So we, who are many, are **one** body in Christ, and individually members **one** of another." Romans 12:5

"Whoever denies the Son does not have the Father; the **one** who confesses the Son has the Father also." 1 John 2:23

"And I give eternal life to them, and they will never perish; and no **one** will snatch them out of My hand." John 10:28

"The glory which You have given Me I have given to them, that they may be **one**, just as we are **one"** John 17:22

Man is either one with Christ, or one with Satan

"Who is the liar but the **one** who denies that Jesus is the Christ? This is the antichrist, the **one** who denies the Father and the Son." 1 John 2:22

"The **one** who listens to you listens to Me, and the **one** who rejects you rejects Me; and he who rejects Me rejects the **One** who sent Me." Luke 10:16

"No **one** can serve two masters; for either he will hate the **one** and love the other, or he will be devoted to **one** and despise the other. You cannot serve God and wealth." Matthew 6:24

"So you will again distinguish between the righteous and the wicked, between **one** who serves God and **one** who does not serve Him." Malachi 3:18

And His flock will be rewarded in oneness with God

"Now in that day the remnant of Israel, and those of the house of Jacob who have escaped, will never again rely on the **one** who struck them, but will truly rely on the Lord, the Holy **One** of Israel." Isaiah 10:20

"But the saints of the Highest **One** will receive the kingdom and possess the kingdom forever, for all ages to come." Daniel 7:18

"And the Lord will be king over all the earth; in that day the Lord will be the only **one**, and His name the only **one**." Zechariah 14:9

We Will Be One With God

The number one symbolizes unity, and God is the Unity of all, for we as His people are in union with God. For as the Lord said "the glory which You have given Me I have given to them, that they may be ***one***, just as We are ***one***". (John 17:22)

And the number one has many unique mathematical properties:

For any number x: x times 1 is equal to x

One is defined as an automorphic number for one is in all things, for one can be placed in any numerical value. And God is automorphic for He is in all things, for He is "***One*** God and Father of all who is over all and through all and in all." (Ephesians 4:6)

For any number x: x divided by x is equal 1

And any number divided by itself is one. And any kingdom, and any house, and even Satan when divided against itself, cannot stand but returns to one. "If a kingdom is divided against itself, that kingdom cannot stand. If a house is divided against itself, that house will not be able to stand. If Satan has risen up against himself and is divided, he cannot stand, but he is finished!"(Mark 3:24-27)

For any number x: x^0 *is equal to 1*

And any number raised to its original value is one. For all were created and shall return to the Father. For "do we not all have ***one*** father? Has not ***one*** God created us?" (Malachi 2:10)

For any power x: 1^x is equal to 1

And the One who is the Father shall transfer power to His people. And the One who is the Son shall raise them up on that day. For as the Son foretold "no ***one*** can come to Me unless the Father who sent Me draws him; and I will raise him up on the last day." (John 6:44)

And one is the atomic number of hydrogen, the atom of the universe which exists in all things. And God is the Creator of the universe, and God the "All in All" exists in all things. For "when all things are subjected to Him, then the Son Himself also will be subjected to the ***One*** who subjected all things to Him, so that God may be all in all."(1 Corinthians 15:28) The word "hydrogen" is derived from the Latin word *hydrogenium* which means "water". And the Father is the water which is the sustenance of all life. "Then he showed me a river of the ***water*** of life, clear as crystal, coming from the throne of God and of the Lamb". (Revelation 22:1) But when man "attempts" to change hydrogen, and when man "attempts" to change God, then the end shall soon come. For when hydrogen is altered then chaos ensues: "And I saw something like a sea of glass mixed with fire." (Revelation 15:2)

Marshall Islands 1946: First Hydrogen Bomb

And the value of alpha in the Greek alphabet is one. And God is the Alpha, the first and the beginning of all things. "I am the ***Alpha*** and the Omega, the first and the last, the beginning and the end." (Revelation 22:13) For there is one God and all men shall soon "know that the Lord is God; there is no ***one*** else." (1 Kings 8:60)

For He is "called the God of all the earth." (Isaiah 54:5) And God is to be feared for "nothing on earth is like Him, ***One*** made without fear". (Job 41:33) And God is "the Mighty ***One***, God, the Lord," who "summoned the earth from the rising of the sun to its setting." (Psalms 50:1) And God is "the ***One*** forming light and creating darkness, causing well-being and creating calamity." (Isaiah 45:7) And God "is the Judge; He puts down one and exalts another." (Psalms 75:7) For "'to whom then will you liken Me that I would be his equal?' says the Holy ***One***." (Isaiah 40:25)

For "***One*** like a Son of Man is coming". (Daniel 7:13-14) For "the ***one*** who sows the good seed is the Son of Man" (Matthew 13:37) For the Son will "go forth for Me to be ruler in Israel." (Micah 5:2) For all things were given to the Son by the Father, and the Son who is the Light reveals the Father to all men. (Matthew 11:27) For as the Son said, "I and the Father are ***one***." (John 10:30) And "the ***One*** who has shone in our hearts to give the Light of the knowledge of the glory of God in the face of Christ." (2 Corinthians 4:6) And now "***One*** is your Father, He who is in heaven" and "***One*** is your Leader, that is, Christ."(Matthew 23:8-10)

And God is at one with us through the Holy Spirit. For as the Lord said of the Spirit "I will give them ***one*** heart, and put a new spirit within them. And I will take the heart of stone out of their flesh and give them a heart of flesh." (Ezekiel 11:19) For "the ***one*** who joins himself to the Lord is ***one*** spirit with Him." (1 Corinthians 6:17) And now "there is ***one*** body and ***one*** Spirit, just as also you were called in ***one*** hope of your calling; ***one*** Lord, ***one*** faith, ***one*** baptism, ***one*** God and Father of all who is over all and through all and in all." (Ephesians 4:4-6) And the day will soon come, when "the Spirit and the bride say, 'Come.' And let the ***one*** who hears say, 'Come.' And let the ***one*** who is thirsty come; let the ***one*** who wishes take the water of life without cost." (Revelation 22:17)

For as the Son promised "the ***one*** who comes to Me I will certainly not cast out." (John 6:37) For His people "who are many, are ***one*** body in Christ, and individually members ***one*** of another." (Romans 12:5) And His people are "***one*** flock with ***one*** shepherd."(John 10:16) For as the Son said "the glory which You have given Me I have given to them, that they may be ***one***, just as we are ***one***." (John 17:22) "And they will never perish; and no ***one*** will snatch them out of His hand." (John 10:28) For the one who confesses to the Son, in the end, the Son shall confess to the Father. (1 John 2:23)

But man remains at enmity with God. For those who are not at one with God, are at one with the ruler of this earth who is Satan. For "who is the liar but the ***one*** who denies that Jesus is the Christ? This is the antichrist, the ***one*** who denies the Father and the Son." (1 John 2:22) And as the Son said to each man of the Father, "the ***one*** who listens to you listens to Me, and the one who rejects you rejects Me; and he who rejects Me rejects the ***One*** who sent Me." (Luke 10:16) For no man can serve both God and mammon, and no man can serve this earth and the One. (Matthew 6:24) And on that day, the Son shall judge between the "***one*** who serves God" and the "***one*** who does not serve Him." (Malachi 3:18)

And on that day His people shall return and "will truly rely on the Lord, the Holy ***One*** of Israel." (Isaiah 10:20) And "the saints of the Highest ***One*** will receive the kingdom and possess the kingdom forever" (Daniel 7:18) "And the Lord will be king over all the earth; in that day the Lord will be the only ***one***, and His name the only ***one***." (Zechariah 14:9)

All is one

αEπ

II

The Duality of Creation

To Andrew the Prophet

Completed November 24, 2007

This world is divided in two

"For this reason a man shall leave his father and his mother, and be joined to his wife; and they shall become one flesh." Genesis 2:24

"God made the **two** great lights, the greater light to govern the day, and the lesser light to govern the night; He made the stars also." Genesis 1:16

"And of every living thing of all flesh, you shall bring **two** of every kind into the ark, to keep them alive with you; they shall be male and female." Genesis 6:19

"For it is written that Abraham had **two** sons, one by the bondwoman and one by the free woman." Galatians 4:22

"The Lord said to her, '**Two** nations are in your womb; and **two** peoples will be separated from your body; and one people shall be stronger than the other; and the older shall serve the younger.'" Genesis 25:23

"'Stop weeping; behold, the Lion that is from the tribe of Judah, the Root of David, has overcome so as to open the book and its seven seals.' And I saw between the throne and the elders a Lamb standing, as if slain, having seven horns and seven eyes, which are the seven Spirits of God, sent out into all the earth." Revelation 5:5-6

God will cut His enemies in two

"The Lord is righteous; He has cut in **two** the cords of the wicked." Psalms 129:4

"No one can serve **two** masters; for either he will hate the one and love the other, or he will be devoted to one and despise the other. You cannot serve God and wealth" Matthew 6:24

"If your eye causes you to stumble, pluck it out and throw it from you. It is better for you to enter life with one eye, than to have **two** eyes and be cast into the fiery hell." Matthew 18:9

"Then there will be **two** men in the field; one will be taken and one will be left. **Two** women will be grinding at the mill; one will be taken and one will be left." Matthew 24:40-41

"And the beast was seized, and with him the false prophet who performed the signs in his presence, by which he deceived those who had received the mark of the beast and those who worshiped his image; these **two** were thrown alive into the lake of fire which burns with brimstone." Revelation 19:20

God's covenants are two

"This is allegorically speaking, for these women are two covenants: one proceeding from Mount Sinai bearing children who are to be slaves; she is Hagar. Now this Hagar is Mount Sinai in Arabia and corresponds to the present Jerusalem, for she is in slavery with her children. But the Jerusalem above is free; she is our mother." Galatians 4:24-26

"It came about at the end of forty days and nights that the Lord gave me the **two** tablets of stone, the tablets of the covenant." Deuteronomy 9:11

"'You shall love the Lord your God with all your heart, your soul, your mind' This is the great and foremost commandment. The second is like it, 'You shall love your neighbor as yourself'. On these **two** commandments depend the whole Law and the Prophets." Matthew 22:37-40

"And behold, the veil of the temple was torn in **two** from top to bottom; and the earth shook and the rocks were split." Matthew 27:51

"By abolishing in His flesh the enmity, which is the Law of commandments contained in ordinances, so that in Himself He might make the **two** into one new man, thus establishing peace" Ephesians 2:15

"And I will make them one nation in the land, on the mountains of Israel; and one king will be king for all of them; and they will no longer be two nations and no longer be divided into **two** kingdoms." Ezekiel 37:22

And Two Shall Be Made One

"In the beginning, God created the heavens and the earth" (Genesis 1:1) The Hebrew word for "in the beginning" is תישאדב. And the first letter of the Torah is *bet* ב , which in Hebrew has the numerical value of two. And the origin of *bet* comes from the Phoenician letter *byt* which gave rise to the Greek letter β. And beta too has the numerical value of two. Thus, the first stroke of the Bible symbolizes the dual nature of His creation, and the dual nature of His Son, and the dual nature of His people, and the dual nature of our works.

And the number of two has many significant mathematical properties:

For from any value of x: $x + x$ *is equal to* $2x$

Thus the value of x is doubled by adding upon oneself to make two. And the nature of our works is symbolized by two, for "whoever forces you to go one mile, go with him ***two***." (Matthew 5:41)

For any value of x: x *times* x *is equal to* x^2

And the value of x when multiplied upon oneself is the square of its root. And His Temple and His City and His Elect, shall be built upon the square of His Root. For the Temple of God is built upon the value of a square, "there shall be for the holy place a ***square*** round about five hundred by five hundred cubits". (Ezekiel 45:2) And the City of God is built upon the value of a square, "the city is laid out as a ***square***, and its length is as great as the width; and he measured the city with the rod, fifteen hundred miles; its length and width and height are equal." (Revelation 21:16) And the number of His Elect is built upon a ***square***, "and I heard the number of those who were sealed, one hundred and forty-four thousand.(twelve tribes by twelve thousand)" (Revelation 7:4)

Two has the unique property of: $2 + 2 = 2 X 2 = 2^2 = 4$

And the addition and the multiplication and the square of two, is equivalent to the value of four. And His Elect is the summation and the multiplication and the square of His Root. For they are the summation of His works, "this very day I am declaring that I will restore ***double*** to you." (Zechariah 9:12) And they are the multiplication of His works, "then I Myself will gather the remnant of My flock out of all the countries where I have driven them and bring them back to their pasture, and they will be fruitful and ***multiply***." (Jeremiah 23:3) And they are the square of His works, "the altar hearth shall be twelve cubits long by twelve wide, ***square*** in its four sides." (Ezekiel 43:16) And they are the product of His Root which is four. For they are the four creatures that worship our Lord, "and the ***four living creatures***, each one of them having six wings, are full of eyes around and within; and day and night they do not cease to say, 'Holy, holy, holy is the Lord God, the Almighty, who was and who is and who is to come.'" (Revelation 4:8)

Two has the unique property of: $\sqrt{2} = 1.414213562 ..... \infty$

For the root of two is an infinite number. And the Root of the Church is an infinite power, for the power is through Jesus, who is the Root of the Church. "I, Jesus, have sent My angel to testify to you these things for the ***churches***. I am the ***root*** and the descendant of David, the bright morning star." (Revelation 22:16)

And in science and in biology, two is of great significant function. For in the process of fertilization, two germ cells are combined and two strands of DNA are intertwined, to form one complete embryo. And a man and his wife are joined to form one flesh. For as the Lord said, "for this reason a man shall leave his father and his mother, and be joined to his wife; and they shall become one flesh." (Genesis 2:24) But the union of the flesh shall soon perish with this earth, and the union of man and wife shall soon be abolished. Then the Bride and the Bridegroom shall form one, and all of the heavens shall rejoice. "Let us rejoice and be glad and give the glory to Him, for the marriage of the Lamb has come and His bride has made herself ready. It was given to her to clothe herself in fine linen, bright and clean; for the fine linen is the righteous acts of the saints." (Revelation 19:7-8)

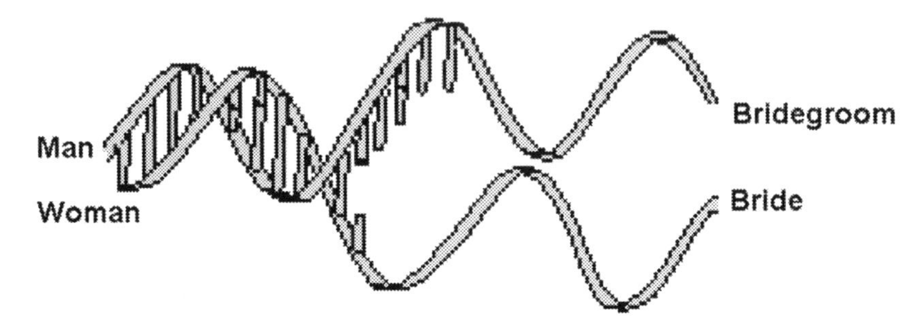

And the state of this world is divided into two. For two must be joined in order to form one flesh. (Genesis 2:24) And the lights of the heavens are divided into two, for the sun lights by day, and the moon lights by night. (Genesis 1:16) And the beasts of the earth were divided into two. (Genesis 6:19) And the nations of Abraham were divided into two, and they shall not be one until all is complete. (Genesis 25:23/Galatians 4:22) And the gate of the Son is divided into two, for He came as the Lamb, but returns as the Lion. (Revelation 5:6)

And the Lion shall return to judge all of mankind, for man is at enmity with God. And His vengeance shall be just for "the Lord is righteous; He has cut in ***two*** the cords of the wicked." (Psalms 129:4) For "no one can serve ***two*** masters; for either he will hate the one and love the other, or he will be devoted to one and despise the other."(Matthew 6:24) For "it is better for you to enter life with one eye, than to have ***two*** eyes and be cast into the fiery hell." (Matthew 18:9) For the day shall soon come when there are two in the field, and "one will be taken and one will be left." (Matthew 24:40-41) And the rulers of this earth shall be cast in the fire, and their rule and their ruler shall soon be abolished. (Revelation 19:20)

And the two covenants of God shall be made into one. For "this is allegorically speaking, for these women are ***two*** covenants: one proceeding from Mount Sinai bearing children who are to be slaves; she is Hagar. Now this Hagar is Mount Sinai in Arabia and corresponds to the present Jerusalem, for she is in slavery with her children. But the Jerusalem above is free; she is our mother." (Galatians 4:24-26) For the covenant of Mount Sinai and Jerusalem was written on two tablets of stone. For "it came about at the end of forty days and nights that the Lord gave me the ***two*** tablets of stone, the tablets of the covenant." (Deuteronomy 9:11) But the covenant of Christ was written on two tablets of flesh, for "you shall love the Lord your God with all your heart your soul your mind" and "you shall love your neighbor as yourself." For "on these ***two*** commandments depends the whole Law and the Prophets." (Matthew 22:37-40) And through Christ's death on the cross, the veil before God was torn into two, and His people are at peace with the Father. (Matthew 27:51) For He has abolished "in His flesh the enmity, which is the Law of commandments contained in ordinances, so that in Himself He might make the ***two*** into one new man, thus establishing peace." (Ephesians 2:15) And when the Son of Man returns, the two kingdoms shall be made into one. "And I will make them one nation in the land, on the mountains of Israel; and one king will be king for all of them; and they will no longer be ***two*** nations and no longer be divided into two kingdoms." (Ezekiel 37:22)

Two shall be made into One

ατπ

III

The Son of Man (NASB)

To Andrew the Prophet

Completed November 27, 2007

The three trinities:

The Crucifixion: The Cross of the Son of Man

3258 BC Father	Son	Holy Spirit
Soul	Mind	Heart
Hope	Love	Faith 2013 AD

"The Son of Man must be delivered into the hands of sinful men, and be crucified, and the **third** day rise again." Luke 24:7

"Go therefore and make disciples of all the nations, baptizing them in the name of the **Father** and the **Son** and the **Holy Spirit**" Matthew 28:19

"And He said to him, 'You shall love the Lord your God with all your **heart**, and with all your **soul**, and with all your **mind**.'" Matthew 22:37

"But now **faith**, **hope**, **love**, abide these **three**; but the greatest of these is love." 1 Corinthians 13:13

Our forefathers were in threes

"On the very same day Noah and Shem and Ham and Japheth, the sons of Noah, and Noah's wife and the three wives of his sons with them, entered the ark. These **three** were the sons of Noah, and from these the whole earth was populated." Genesis 7:13, 19

"Suddenly the Lord said to Moses and Aaron and to Miriam, 'You **three** come out to the tent of meeting.'" Numbers 12:4

"And if one can overpower him who is alone, two can resist him. A cord of **three** strands is not quickly torn apart." Ecclesiastes 4:12

And the days of waiting are three

"for just as Jonah was **three** days and three nights in the belly of the sea monster , so will the Son of Man be three days and three nights in the heart of the earth." Matthew 12:40

"and said, 'This man stated, 'I am able to destroy the temple of God and to rebuild it in **three** days.'" Matthew 26:61

"Those from the peoples and tribes and tongues and nations will look at their dead bodies for **three** and a half days, and will not permit their dead bodies to be laid in a tomb." Revelations 11:9

And the Holy City is in three

"You shall set aside **three** cities for yourself in the midst of your land, which the Lord your God gives you to possess." Deuteronomy 19:2

"The great city was split into **three** parts, and the cities of the nations fell. Babylon the great was remembered before God, to give her the cup of the wine of His fierce wrath." Revelations 16:19

"There were **three** gates on the east and **three** gates on the north and **three** gates on the south and **three** gates on the west." Revelations 21:13

The Trinity of God of Man and of His Graces

"Go therefore and make disciples of all the nations, baptizing them in the name of the Father and the Son and the Holy Spirit." (Matthew 28:19) The time is near but the calling remains the same. For God the Trinity has returned for all of mankind. For the Father is the Water and the sustenance of all life; the Son is the Word and the Blood for mankind; and the Spirit is the Heart which yields the Fruit for His kingdom. For God is three in one, for God is the Holy Trinity. And the concept that the Trinity is "three in one" is not as difficult as it may seem. For "God created man in His own image, in the image of God He created him." (Genesis 1:27) For just as God is three in one, so is man created three in one. For each man is a trinity having "three in one": a soul which gives an inheritance with the Father; a spirit which gives the heart to accomplish His works; and a mind which gives the Word and the knowledge of God. And through the baptism of the Holy Spirit, we receive the triune of graces from God: hope - that we may return to the Father; faith - that through the Spirit we may bear His fruits; and love - that we may sacrifice our lives for Christ, even unto death. (1 Corinthians 13:13) And through the three graces of the Spirit, the Kingdom has been leavened, and has grown to fruition for all of mankind. "The kingdom of heaven is like leaven, which a woman took and hid in ***<u>three</u>*** pecks of flour until it was all leavened." (Matthew 13:33)

And three has several important mathematical properties:

Three is the lowest prime number which cannot be reciprocated into a decimal value. In other words if 1 is divided into 3, the number becomes undefined. i.e. 0.3333333333333...∞. And just as one cannot be divided into three, the Trinity cannot be divided into three, nor can man's nature be divided into three, nor can God's graces be divided into three. For they are all intrinsically one. "For there are ***three*** that testify: the Spirit and the water and the blood; and the ***three*** are in agreement." (1 John 5:7-8)

Three also symbolizes a triangle, which is a structure of great strength. For a rectangle topples over if its angles are not fixed. But a triangle is a structure whose shape never changes. And the Trinity is the foundation of creation, and this rock and this refuge shall never change. "On God my salvation and my glory rest; the rock of my strength, my refuge is in God." (Psalms 62:7)

The Lord is my Rock

And three is a number of great significance in science:

Light consists of three primary colors: red, green, and blue. And when these colors are combined, the pigmented colors of the earth turn dark, but the primary colors of light become white. And the Son who is the Light, turns darkness into Light, for as He said "I am the Light of the world; he who follows Me will not walk in the darkness, but will have the Light of life." (John 8:12) And now His children are Light and like the Light, the sons of all nations will be adorned in robes of white. "These who are clothed in the white robes, who are they, and where have they come from?' I said to him, 'My lord, you know.' And he said to me, 'These are the ones who come out of the great tribulation, and they have washed their robes and made them white in the blood of the Lamb. For this reason, they are before the throne of God; and they serve Him day and night in His temple; and He who sits on the throne will spread His tabernacle over them.'" (Revelation 7:13-15)

And three is the atomic number of lithium. For lithium was one of the first atoms to be split, and is now used as fuel for nuclear weapons. And as was foretold, "the first trumpet sounded, and there came hail and fire, mixed with blood, and they were thrown to the earth; and a third of the earth was burned up, and a third of the trees were burned up, and all the green grass was burned up." (Revelation 8:7) And ironically, the first nuclear bomb was detonated in Los Alamos, and it was nicknamed the "Trinity Bomb".

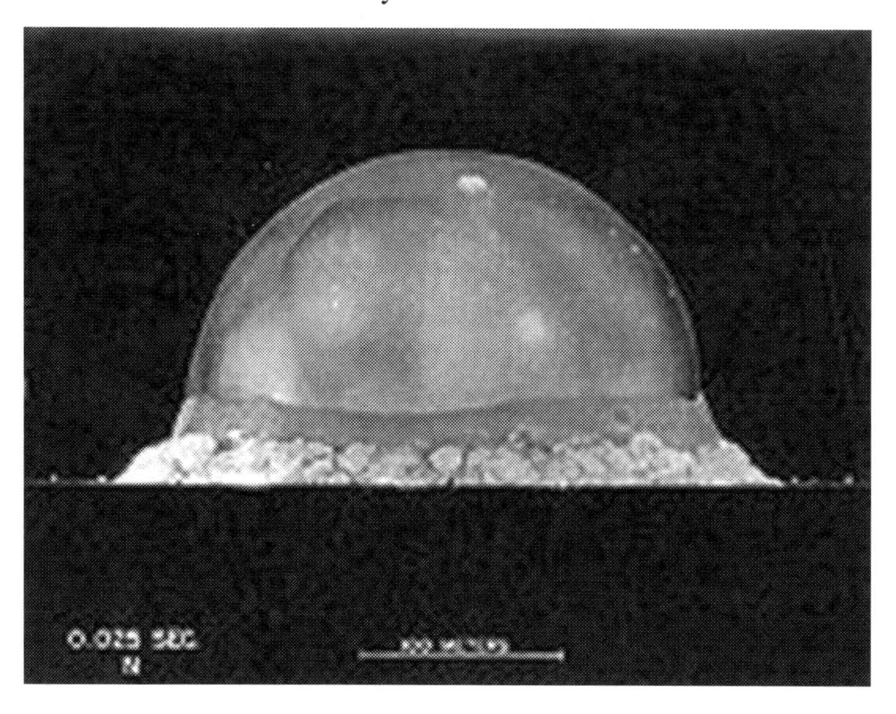

The First Stages of the Trinity Bomb

And three is represented in Roman numerals by III: the three pillars that hold up the temple of God. "So the great court all around had ***three*** rows of cut stone." (1 King 7:12) For His Elect shall become the pillar of His temple, for as He said "he who overcomes, I will make him a pillar in the temple of My God, and he will not go out from it anymore; and I will write on him the name of My God, and the name of the city of My God, the new Jerusalem, which comes down out of heaven from My God, and My new name." (Revelation 3:12) And the symbol of His people is three. For the star of David is the summation of two triangles. For when all things are turned over to the King, the triangle will be turned over and the star will be complete. "He who overcomes, and he who keeps My deeds until the end, to him I will give authority over the nations; and he shall rule them with a rod of iron, as the vessels of the potter are broken to pieces, as I also have received authority from My Father; and I will give him the ***morning star***." (Revelation 2:26-28)

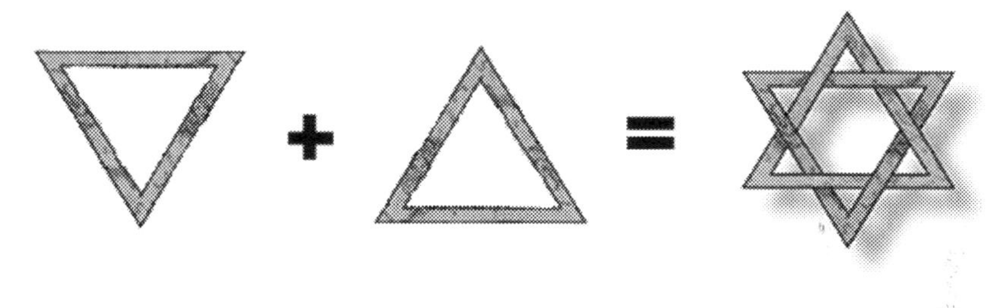

And the history of our forefathers is in threes. For all men are descendants of the three sons of Noah. (Genesis 7:13, 19) And the three leaders delivered the Israelites through the sea. (Numbers 12:4) For though the oppression of Satan is great, the union of three is stronger than he; for "a cord of ***three*** strands is not quickly torn apart." (Ecclesiastes 4:12)

And the symbol of 3 when turned upside down becomes ε, which symbolizes the end of time. For "this gospel of the kingdom shall be preached in the whole world as a testimony to all the nations, and then the end will come." (Matthew 24:14) And the symbol of 3 when turned upside down is € for Euros, the cause for entering the end of all wars. (Addendum 3) For this reason, the war was started for oil and for currency, and for wine and for blood. For OPEC was threatening to exchange oil-currency in Euros, thus threatening the economic strength of the US dollar. Thus Bush and his cronies went to war with Iraq, and confiscated their oil and poured out their blood. But the third seal shall be opened in 2009, and the sons of Satan will pay for oil with their blood. For as was foretold "a quart of wheat for a denarius, and ***three*** quarts of barley for a denarius; and do not damage the oil and the wine." (Revelation 6:6)

And the time of waiting is for three days and nights. And for three days and nights, just as Jonah descended into the belly of a whale, so our Lord descended into the belly of hell, to set its captives free. (Matthew 12:40) For by the sacrifice of His body and His blood, the temple of this earth was destroyed, and exchanged for the Temple in heaven. (Matthew 26:61) And His witness shall perish in Jerusalem, at the site of the death of our Lord. And his body shall lie in Jerusalem, but in three days and nights, he shall arise to proclaim the return of the King! (Revelation 11:9)

And His city shall be established in three. For when His people entered the holy land, He instructed them to build three cities, and the holiest of these three cities was Jerusalem. (Deuteronomy 19:2) And when the King returns and the second woe calls, Jerusalem shall be split into three. (Revelation 16:19) And His people shall be raised up in chariots, and the Temple of God shall be opened. (Ezekiel 40) And the new Jerusalem shall be seen for all to behold. And its walls shall have three gates on each side, so that all who are worthy may enter. (Revelation 21:13) "I saw no temple in it, for the Lord God the Almighty and the Lamb are its temple. And the city has no need of the sun or of the moon to shine on it, for the glory of God has illumined it, and its lamp is the Lamb. The nations will walk by its light, and the kings of the earth will bring their glory into it. In the daytime (for there will be no night there) its gates will never be closed; and they will bring the glory and the honor of the nations into it; and nothing unclean, and no one who practices abomination and lying, shall ever come into it, but only those whose names are written in the Lamb's book of life." (Revelation 21:22-27)

IV

A Time Four Change

To Andrew the Prophet

Completed November 29, 2007

Revelation 4: "And the **four** living creatures, each one of them having six wings, are full of eyes around and within; and day and night they do not cease to say, 'Holy, holy, holy is the Lord God, the Almighty, who was and who is and who is to come.'"

The return of God is represented by four

"God said to Moses, 'I AM WHO I AM' [Related to the name of God, **YHWH**, rendered **LORD**]" Exodus 3:14

"And each one had **four** faces. The first face was the face of a cherub, the second face was the face of a man, the third the face of a lion, and the fourth the face of an eagle." Ezekiel 10:14

The remnant of Israel is in four

"And He will lift up a standard for the nations and assemble the banished ones of Israel, And will gather the dispersed of Judah From the **four** corners of the earth." Isaiah 11:12

"Now the altar hearth shall be twelve cubits long by twelve wide, square in its **four** sides." Ezekiel 43:16

"The first creature was like a lion, and the second creature like a calf, and the third creature had a face like that of a man, and the **fourth** creature was like a flying eagle." Revelations 4:7

"And they sang a new song before the throne and before the **four** living creatures and the elders; and no one could learn the song except the one hundred and **forty-four** thousand who had been purchased from the earth." Revelations 14:3

The four beasts and four horns

“and exchanged the glory of the incorruptible God for an image in the form of corruptible man and of birds and **four-footed** animals and crawling creatures.” Romans 1:23

“And **four** great beasts were coming up from the sea, different from one another. The first was like a lion and had the wings of an eagle. I kept looking until its wings were plucked, and it was lifted up from the ground and made to stand on two feet like a man; a human mind also was given to it. And behold, another beast, a second one, resembling a bear. And it was raised up on one side, and three ribs were in its mouth between its teeth; and thus they said to it, ‘Arise, devour much meat!’ After this I kept looking, and behold, another one, like a leopard, which had on its back **four** wings of a bird; the beast also had **four** heads, and dominion was given to it. After this I kept looking in the night visions, and behold, a fourth beast, dreadful and terrifying and extremely strong; and it had large iron teeth. It devoured and crushed and trampled down the remainder with its feet; and it was different from all the beasts that were before it, and it had ten horns.8 While I was contemplating the horns, behold, another horn, a little one, came up among them, and three of the first horns were pulled out by the roots before it; and behold, this horn possessed eyes like the eyes of a man and a mouth uttering great boasts.” Daniel 7:3-8

His four angels will be released

“Now I lifted up my eyes again and looked, and behold, **four** chariots were coming forth from between the two mountains; and the mountains were bronze mountains. The angel replied to me, ‘These are the **four** spirits of heaven, going forth after standing before the Lord of all the earth.’” Zechariah 6:1, 5

After this I saw **four** angels standing at the **four** corners of the earth, holding back the **four** winds of the earth, so that no wind would blow on the earth or on the sea or on any tree. And I saw another angel ascending from the rising of the sun, having the seal of the living God; and he cried out with a loud voice to the **four** angels to whom it was granted to harm the earth and the sea, saying, 'Do not harm the earth or the sea or the trees until we have sealed the bond-servants of our God on their foreheads.'" Revelation 7:1-3

"And then He will send forth the angels, and will gather together His elect from the **four** winds, from the farthest end of the earth to the farthest end of heaven." Mark 13:27

"one saying to the sixth angel who had the trumpet, 'Release the **four** angels who are bound at the great river Euphrates.' And the **four** angels, who had been prepared for the hour and day and month and year, were released, so that they would kill a third of mankind." Revelations 9:14-15

Time the Last Ruler Shall Be Destroyed

"In a moment, in the twinkling of an eye, at the last trumpet; for the trumpet will sound, and the dead will be raised imperishable, and we will be changed." (1 Corinthians 15:52) And what is the greatest curse upon mankind? It is not the fruit of the garden. For the greatest curse is that which imprisons our lives, for our lives are limited by dust which is death. Thus the greatest curse upon mankind is time. For by the curse of time, Satan has held dominion over man's life. For as is said "cursed is the ground because of you; in toil you will eat of it. All the days of your life." (Genesis 3:17) But the time has come for change, and the toil of this earth shall soon end. For the curse of time has been turned against Satan, and now time shall hold its curse over his rule. And now he has been cast down to earth, but "woe to the earth and the sea, because the devil has come down to you, having great wrath, knowing that he has only a short time." (Revelation 12:12)

And the time for change is now.

For the number of four is the Greek letter delta (**Δ**). And a delta is a portion of land seated at the mouth of a river. And the Garden of Eden was seated at the delta of four rivers. "Now a river flowed out of Eden to water the garden; and from there it divided and became ***four*** rivers." (Genesis 2:10) And scientifically delta represents change, for a great change will occur on that day. "And behold, some are last who will be first and some are first who will be last." (Luke 13:30) And the delta shall be turned over to God, and the star of His people shall be complete. For four symbolizes a change in the ruler of this earth, of the remnant of His people, of the tolerance of injustice, and of the dominion of time.

Δ + ∇ = ✡

And the return of the King is in four. For the name of our Lord is four letters. For He is called "I AM WHO I AM", for He is called ה ו ה י, for He is called "Y H W H", for He is called "L O R D". (Exodus 3:14) And His name was seen in the creature with four faces, "And each one had ***four*** faces. The first face was the face of a cherub, the second face was the face of a man, the third the face of a lion, and the fourth the face of an eagle." (Ezekiel 10:14) The first creature is a cherub, for this is the Son of God, who came as a child to show us the Light. And the second creature is the face of a man, for this is the face of the Father, who sent His only Son to teach us the Word. And the third creature is the face of a Lion, for this is Christ the Lion, who comes as the Lion to judge all mankind. And the fourth creature is the face of an eagle, for this is the Spirit of God, who descended to mankind by Christ's death on the cross.

And the remnant of His people is in four. For the prophets of old foretold, that the remnant and the banished would return, from "the ***four*** corners of the earth." (Isaiah 11:12) And the hearth of the altar of His temple, is the sacrifice of the remnant of His people. "The altar hearth shall be 12 cubits long by 12 wide, square in its ***four*** sides." (Ezekiel 43:16) For His remnant are the prophets and martyrs and saints, for they shall be 12 tribes long by 12 thousand wide. (Revelation 7:5-8) For they were created in the image of God, for His remnant is the creature with four heads, which worships the Lord both day and night. And the remnant and the heavenly hosts shall sing "Holy, holy, holy is the Lord God, the Almighty, who was and who is and who is to come." (Revelation 4:7-8) And "no one could learn the song except the one hundred and ***forty-four*** thousand who had been purchased from the earth." (Revelation 14:3)

But woe to the world for they are at enmity with God. For they have forsaken the Lord, "and exchanged the glory of the incorruptible God for an image in the form of corruptible man and of birds and ***four***-footed animals and crawling creatures." (Romans 1:23) And Daniel had prophesied that the four beasts would return. For the time has arrived, for the four beasts of perdition to destroy all of mankind. For the four beasts of perdition, are Bush, Cheney, al-Sadr, and the nations, who will destroy mankind through their hatred and deception. (Daniel 7:3)

"The first was like a lion and had the wings of an eagle. I kept looking until its wings were plucked, and it was lifted up from the ground and made to stand on two feet like a man; a human mind also was given to it." (Daniel 7:4) For the lion with the wings of an eagle is George Bush Jr. He comes from the US across the great ocean, to war with Iraq for their currency and their oil. For the eagle with two wings represents his executive power: one wing represents military power, and the other wing represents economic strength. But because of his haughtiness and his God-hating ways, his wings are finally torn from him. And because he could not think like a man, now he must stand up like a man

"And behold, another beast, a second one, resembling a bear. And it was raised up on one side, and three ribs were in its mouth between its teeth; and thus they said to it, 'Arise, devour much meat!'" (Daniel 7:5) For the beast and the bear is Dick Cheney. For the bear has feasted on and devoured three nations. For Iraq, Afghanistan, and Kuwait are the nations he destroyed by his corruption and his greed.

"After this I kept looking, and behold, another one, like a leopard, which had on its back four wings of a bird; the beast also had four heads, and dominion was given to it." (Daniel 7:6) For the beast is Moqtada al-Sadr, and like a leopard kills with great cunning. He will unify the nations through stealth and deceit. And the four heads are the nations he will unify: Iraq, Iran, Pakistan, and Libya. And at the end they will battle against the fourth beast. And the four wings are his power and his authority, to unify the nations despite their great fall.

"After this I kept looking in the night visions, and behold, a fourth beast, dreadful and terrifying and extremely strong; and it had large iron teeth. It devoured and crushed and trampled down the remainder with its feet; and it was different from all the beasts that were before it, and it had ten horns." (Daniel 7:7) For the fourth beast is the union of ten nations, that will battle the beast with four heads. And the ten nations are the United States, Canada, England, Germany, Australia, New Zealand, South Korea, Russia, China, and Saudi Arabia. And their iron teeth signify that their strength is like iron. For their military strength shall be great, and no other nation shall stand against them...that is except with nuclear weapons.

"While I was contemplating the horns, behold, another horn, a little one, came up among them, and three of the first horns were pulled out by the roots before it; and behold, this horn possessed eyes like the eyes of a man and a mouth uttering great boasts." (Daniel 7:8) For the little horn is Moqtada al-Sadr, and through his craftiness and deceit, he will overtake the other three horns. And from his mouth he utters great boasts, "No, no Satan, No, no USA, No, no occupation. No, no Israel" (AP 5/2007)

"And behold, this horn possessed eyes" Daniel 7:8

Then the four angels who watch over the earth shall be released. For man has ignored their power and their strength. For Michael and Gabriel and Raphael and Uriel "are the ***four*** spirits of heaven, going forth after standing before the Lord of all the earth." (Zechariah 6:1,5) And they keep account of every transgression, for as the Lord promised "but I tell you that every careless word that people speak, they shall give an accounting for it in the day of judgment." (Matthew 12:36) And even now during the unfolding of these times, the four patiently wait as the Lord has instructed them, "do not harm the earth or the sea or the trees until we have sealed the bond-servants of our God on their foreheads." (Revelation 7:1-3) And they shall go forth throughout all the earth, to "gather together His elect from the ***four*** winds, from the farthest end of the earth to the farthest end of heaven." (Mark 13:27) But woe to those who are left behind! For the trumpet will sound, then God will "'release the ***four*** angels who are bound at the great river Euphrates.' And the

four angels, who had been prepared for the hour and day and month and year, were released, so that they would kill a third of mankind." (Revelation 9:14-15)

For the time to prepare is now.

For the fourth dimension in physics is the continuum known as time. For time has held its grip on the earth, and by its power it has enslaved mankind. But His saints have patiently awaited for this time, "How long, O Lord, holy and true, will You refrain from judging and avenging our blood on those who dwell on the earth?" (Revelation 6:10) For His saints have patiently waited as instructed: "be patient, brethren, until the coming of the Lord. The farmer waits for the precious produce of the soil, being patient about it, until it gets the early and late rains. You too be patient; strengthen your hearts, for the coming of the Lord is near." (James 5:7-8) For their patience shall be greatly rewarded, for the King shall return in the clouds, for the time of the judgment is near, and time shall be vanquished, and time shall be no more.

TIME

V

Five: The Hand of God (NKJV)

To Andrew the Prophet

Completed January 8, 2008

The Tabernacle was made in the image of God's hands

"**Five** curtains shall be joined to one another, and the other curtains shall be joined to one another." Exodus 26:3

"Then you shall make bars of acacia wood, **five** for the boards of one side of the tabernacle, and **five** bars for the boards of the other side of the tabernacle, and **five** bars for the boards of the side of the tabernacle for the rear side to the west." Exodus 26:26-27

"You shall make **five** pillars of acacia for the screen and overlay them with gold, their hooks also being of gold; and you shall cast **five** sockets of bronze for them." Exodus 26:37

The temple of God was made in the semblance of God's hands

"**Five** cubits was the one wing of the cherub and **five** cubits the other wing of the cherub; from the end of one wing to the end of the other wing were ten cubits." 1 Kings 6:24

"He also made ten basins in which to wash, and he set **five** on the right side and **five** on the left to rinse things for the burnt offering; but the sea was for the priests to wash in. Then he made the ten golden lampstands in the way prescribed for them and he set them in the temple, **five** on the right side and **five** on the left. He also made ten tables and placed them in the temple, **five** on the right side and **five** on the left. And he made one hundred golden bowls." 2 Chronicles 4:6-8

Christ is the Hands of God

"Ordering the people to sit down on the grass, He took the **five** loaves and the two fish, and looking up toward heaven, He blessed the food, and breaking the loaves He gave them to the disciples, and the disciples gave them to the crowds." Matthew 14:19

"There were about **five** thousand men who ate, besides women and children." Matthew 14:21

"when I broke the **five** loaves for the **five** thousand, how many baskets full of broken pieces you picked up?" They said to Him, 'Twelve.'" Mark 8:19

His followers represent the hands of God

"A certain nobleman went into a far country to receive for himself a kingdom and to return. So he called ten of his servants, delivered to them ten minas, and said to them, 'Do business till I come.' But his citizens hated him, and sent a delegation after him, saying, 'We will not have this man to reign over us.' And so it was that when he returned, having received the kingdom, he then commanded these servants, to whom he had given the money, to be called to him, that he might know how much every man had gained by trading. Then came the first, saying, 'Master, your mina has earned ten minas.' And he said to him, 'Well done, good servant; because you were faithful in a very little, have authority over ten cities.' And the second came, saying, 'Master, your mina has earned **five** minas.' Likewise he said to him, 'You also be over **five** cities.' Then another came, saying, 'Master, here is your mina, which I have kept put away in a handkerchief. For I feared you, because you are an austere man. You collect what you did not deposit, and reap what you did not sow.' And he said to him, 'Out of your own mouth I will judge you, you wicked servant. You knew that I was an austere man, collecting what I did not deposit and reaping what I did not sow. Why then did you not put my money in the bank, that at my coming I might have collected it with interest?' And he said to those who stood by, 'Take the mina from him, and give it to him who has ten minas.' (But they said to him, 'Master, he has ten minas.') For I say to you, that to everyone who has will be given; and from him who does not have, even what he has will be taken away from him. But bring here those enemies of mine, who did not want me to reign over them, and slay them before me." Luke 19:12-27

The name of His Word is five letters

"However, in the church I desire to speak **five** words with my mind so that I may instruct others also, rather than ten thousand words in a tongue." 1 Corinthians 14:19

"Jesus Christ, Son of God Savior; Iesous Christos Theou Yios Soter." ΙΧΘΥΣ

The punishment is famine

"Then the **fifth** angel sounded, and I saw a star from heaven which had fallen to the earth; and the key of the bottomless pit was given to him." Revelation 9:1

"And they were not permitted to kill anyone, but to torment for **five** months; and their torment was like the torment of a scorpion when it stings a man." Revelation 9:5

"They have tails like scorpions, and stings; and in their tails is their power to hurt men for **five** months." Revelation 9:10

"Then the **fifth** angel poured out his bowl on the throne of the beast, and his kingdom became darkened; and they gnawed their tongues because of pain." Revelation 16:10

Creazione di Adamo
Michelangelo

The Hand of God Touches Man

It is by the hand of God that He shows His love and kindness, and it is by the hand of God that in vengeance He destroys. "Show Your marvelous lovingkindness by Your right hand, O You who save those who trust in You from those who rise up against them." (Psalms 17:7) For God controls all things in His hands, and He created man from the work of His hands. For the Roman numeral for five is V, which is the creative power of His hands. And the Greek numeral for five is epsilon (ε), which is the yearning of life from our hands. For "You open Your hand and satisfy the desire of every living thing." (Psalms 145:16)

Hand of God

Hand of Man

And the Word came from the hands of the Father. For the covenant of the Ten Commandments, was written by His hands. "And when He had made an end of speaking with him on Mount Sinai, He gave Moses two tablets of the Testimony, tablets of stone, written with the finger of God." (Exodus 31:18) And the Torah are the five books of His people, which form the foundation of His Word.

And the tabernacle was built in the semblance of His hands. For by the hand of God, His people were delivered from Egypt. "By strength of hand the Lord brought us out of Egypt, out of the house of bondage." (Exodus 13:14) And by the strength of His hands, the tabernacle of the covenant was built. For the tabernacle was covered by five curtains on each side. (Exodus 26:3) And the tabernacle was built with five boards on each sides. (Exodus 26:26-27) And the tabernacle was held by five pillars on each side. (Exodus 26:37)

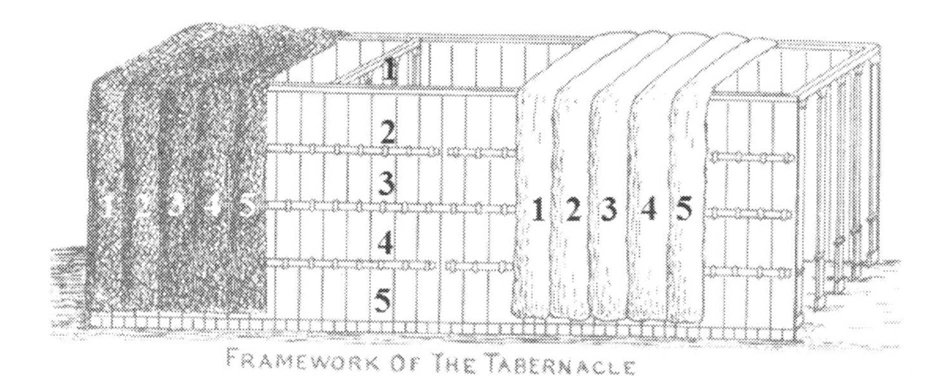

FRAMEWORK OF THE TABERNACLE

And the temple was built in the semblance of His hands. For the cherubim was the breadth of five cubits on each side. (1 Kings 6:24) And in the holy place were five basins on each side, on which the priests cleaned the Lord's offering. And in front of the basins in the temple, were five tables of sacrifice on each side, on which to prepare the Lord's offering. And besides the basins of sacrifice, were five lampstands of gold on each side, on which to light the Lord's offering. And on each table of offering in the temple, were five bowls of gold on each side, on which to present the Lord's offering. (2 Chronicles 4:6-8) And the priests were to keep their hands pure. For "he who has clean hands and a pure heart, who has not lifted up his soul to an idol, nor sworn deceitfully. He shall receive blessing from the Lord." (Psalms 24:4-5) But His priests did not keep their hands pure, thus the temple of their hands was destroyed. (Ezra 5:12)

Thus the Son came as the High Priest of the Temple. And the High Priest of the Temple "took the ***five*** loaves and the two fish, and looking up toward heaven, He blessed the food, and breaking the loaves He gave them to the disciples, and the disciples gave them to the crowds." (Matthew 14:19) For by the power of His hands, from five loaves the Priest would feed five thousand. And from the hands of the Temple, the priesthood was handed to His disciples. And from the Word of the High Priest, one priest would give bread to one thousand. (Mark 8:19) And now He has handed the priesthood to His people. And each has been called as a priest for His kingdom.

For some will be handed one, and some will be handed five, and some will be handed ten. And each will be rewarded for their hand in His kingdom. But for those who have forsaken His name, their inheritance shall be taken and handed to His servants. (Luke 19:12-27) And that which is left shall be thrown in the fire. (Luke 3:17)

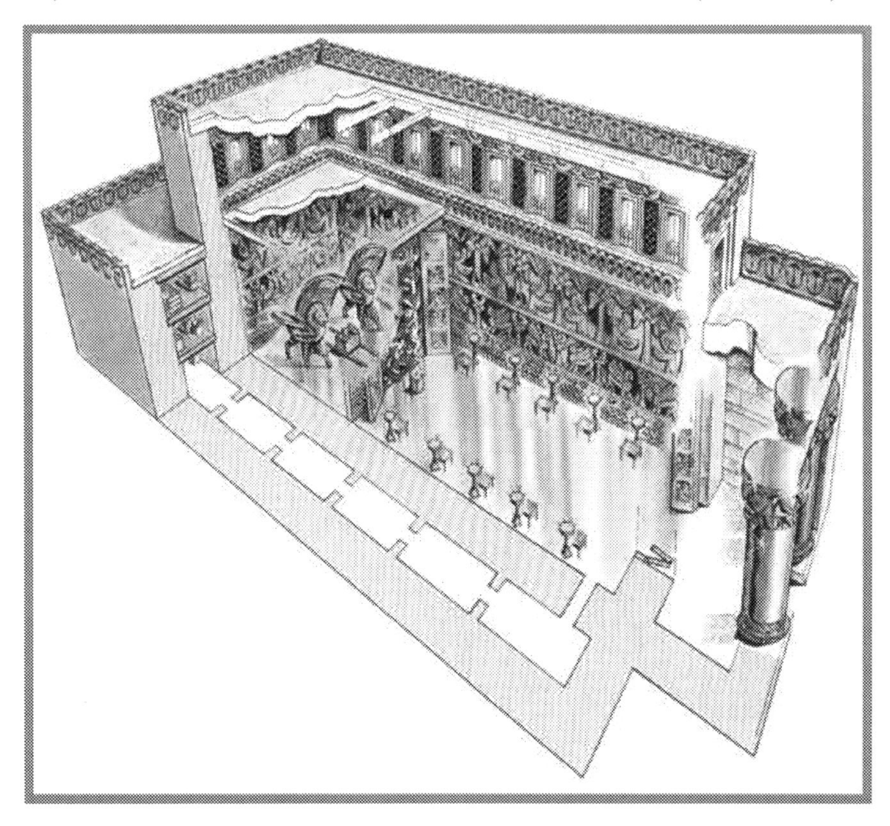

And there are five words which advance the Word and the Truth. For these five words are "Jesus Christ, Son of God Savior: Iesous Christos Theou Yios Soter." For Jesus Christ is the Son of God, who gave us the Word and the Truth, in order to save His people from death. And His people have been handed the priesthood, and have spread the Word and the Truth. For the five words of His name is **Ι Χ Θ Υ Σ** or I C T Y S, which in Latin is translated to "fish". For as He said, "follow Me, and I will make you become fishers of men." (Mark 1:17)

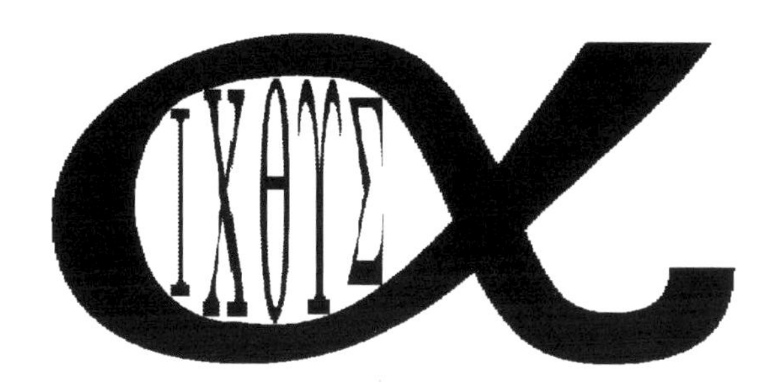

But just as God's hand created all mankind, so shall His hand destroy all mankind. For the fifth trumpet shall warn of the terror of His judgment. (Revelation 9:1) And when the fifth bowl is poured out, a nuclear winter shall cover the earth. (Revelation 16:10) And the earth shall be covered in darkness, and the land and all sea life shall perish. And mankind will suffer from pestilence and famine, for"they will gnaw on their tongues because of pain." (Revelation 9:5; 16:10)

So remember the power of God's righteous hand, for the hand of His judgment is near. For His right hand shall redeem all His people. But do not plead the "fifth", for His right hand shall also destroy what is left! "The Lord is at Your right hand; He shall execute kings in the day of His wrath. He shall judge among the nations, He shall fill the places with dead bodies, He shall execute the heads of many countries. He shall drink of the brook by the wayside; therefore He shall lift up the head." (Psalms 110:5-7)

VI

Six: One Step Away From Perfection (NKJV)

To Andrew the Prophet

Completed January 11, 2008

God created man on the sixth day

"Then God said, 'Let Us make man in Our image, according to Our likeness; let them have dominion over the fish of the sea, over the birds of the air, and over the cattle, over all the earth and over every creeping thing that creeps on the earth.' So God created man in His own image; in the image of God He created him; male and female He created them. Then God blessed them, and God said to them, 'Be fruitful and multiply; fill the earth and subdue it; have dominion over the fish of the sea, over the birds of the air, and over every living thing that moves on the earth.' And God said, 'See, I have given you every herb that yields seed which is on the face of all the earth, and every tree whose fruit yields seed; to you it shall be for food. Also, to every beast of the earth, to every bird of the air, and to everything that creeps on the earth, in which there is life, I have given every green herb for food'; and it was so. Then God saw everything that He had made, and indeed it was very good. So the evening and the morning were the **sixth** day." Genesis 1:26-31

Man's work is completed in six

"For in **six** days the Lord made the heavens and the earth, the sea, and all that is in them, and rested the seventh day. Therefore the Lord blessed the Sabbath day and hallowed it." Exodus 20:11

"**Six** days you shall gather it, but on the seventh day, which is the Sabbath, there will be none." Exodus 16:26

"**Six** years you shall sow your land and gather in its produce" Exodus 23:10

"**Six** days you shall do your work, and on the seventh day you shall rest, that your ox and your donkey may rest, and the son of your female servant and the stranger may be refreshed." Exodus 23:12

"If you buy a Hebrew servant, he shall serve **six** years; and in the seventh he shall go out free and pay nothing." Exodus 21:2

God's tabernacle was created in six

"And **six** branches shall come out of its sides: three branches of the lampstand out of one side, and three branches of the lampstand out of the other side." Exodus 25:32

"For the far side of the tabernacle, westward, you shall make **six** boards." Exodus 26:22

"**six** of their names on one stone, and **six** names on the other stone, in order of their birth." Exodus 28:10

"You shall set them in two rows, **six** in a row, on the pure gold table before the Lord" Leviticus 24:6

God's temple was created in six

"Thus says the Lord God: 'The gateway of the inner court that faces toward the east shall be shut the **six** working days; but on the Sabbath it shall be opened, and on the day of the New Moon it shall be opened.'" Ezekiel 46:1

"The throne had **six** steps, and the top of the throne was round at the back; there were armrests on either side of the place of the seat, and two lions stood beside the armrests." 1 Kings 10:19

"Twelve lions stood there, one on each side of the **six** steps; nothing like this had been made for any other kingdom." 2 Chronicles 9:19

God perfected His servants in six

"He shall deliver you in **six** troubles, Yes, in seven no evil shall touch you." Job 5:19

"the glory of the Lord rested on Mount Sinai, and the cloud covered it **six** days. And on the seventh day He called to Moses out of the midst of the cloud." Exodus 24:16

"You shall march around the city, all you men of war; you shall go all around the city once. This you shall do **six** days." Joshua 6:3

"Now after **six** days Jesus took Peter, James, and John, and led them up on a high mountain apart by themselves; and He was transfigured before them." Mark 9:2

"But the ruler of the synagogue answered with indignation, because Jesus had healed on the Sabbath; and he said to the crowd, 'There are **six** days on which men ought to work; therefore come and be healed on them, and not on the Sabbath day.'" Luke 13:14

The worst of mankind shall be judged in the sixth bowl

"I looked when He opened the **sixth** seal, and behold, there was a great earthquake; and the sun became black as sackcloth of hair, and the moon became like blood." Revelations 6:12

"Then the **sixth** angel sounded: And I heard a voice from the four horns of the golden altar which is before God saying to the sixth angel who had the trumpet, 'Release the four angels who are bound at the great river Euphrates.'" Revelations 9:14

"Then the **sixth** angel poured out his bowl on the great river Euphrates, and its water was dried up, so that the way of the kings from the east might be prepared." Revelations 16:12

The Sixth Element Shall Burn in the Fire

On the sixth day God created Adam and Eve. And "God created man in His own image, in the image of God He created him; male and female." (Genesis 1:27) And from the seed of Adam and Eve, the fruit of the womb has spread through the earth. (Genesis 1:26-31) For man was created in the image of God, but due to man's fall he is dust of the earth. "Because from it you were taken; for you are dust, and to dust you shall return." (Genesis 3:19) And the dust of the earth, shall burn like ember in the fire. For the French word for ember is *charbone,* which in Latin is translated to *carbo.* And the element of carbon (C) has the atomic number six. And the Greek letter for six is *stigma* (ς). Thus carbon is the stigma of man's fall from God's grace.

Carbon is an essential element of the earth, for the life cycle of the earth is bounded by this element. For "even so we, when we were children, were in bondage under the elements." (Galatians 4:3) And the element of carbon is diverse, for all organic compounds are formed by its structure. For carbon can be formed into a beautiful diamond, through which the sun can brilliantly shine. And the firstfruits of God are like diamonds in His hands. For their "whole body is full of light, with no dark part in it, it will be wholly illumined, as when the lamp illumines you with its rays."(Luke 11:36) "Let your light shall shine before men, that they may see your good works and glorify your Father in heaven." (Matthew 5:16)

Yet carbon can be degraded into oil, which is dark as the night and burns fervently in fire. And the sons of perdition are as dark as their ruler, for they have refused to follow the Light. And "this is the judgment, that the Light has come into the world, and men loved the darkness rather than the Light, for their deeds were evil. For everyone who does evil hates the Light, and does not come to the Light for fear that his deeds will be exposed." (John 3:19-20) But "the Lord comes who will both bring to light the things hidden in the darkness and disclose the motives of men's hearts." (1 Corinthians 4:5) And "the day of the Lord will come as a thief in the night, in which the heavens will pass away with a great noise, and the elements will melt with fervent heat; both the earth and the works that are in it will be burned up." (2 Peter 3:10)

And the recycling of carbon is through the carbon cycle. And the most common compound in the atmosphere is carbon dioxide (CO_2). It is converted by plants through photosynthesis into carbohydrate ($C_6H_{12}O_6$). And carbohydrate is consumed by animals and by man, and through the process of cellular respiration, is converted back into CO_2 and water ($H_2 0$). Thus the overall pathway of the carbon cycle is:

PHOTOSYNTHESIS:

$$6\ CO_2 + LIGHT + 6\ H_20 \rightarrow C_6H_{12}O_6 + 6\ O_2$$

CELLULAR RESPIRATION:

$$C_6H_{12}O_6 + 6\ O_2 \rightarrow 6\ CO_2 + 6\ H_20 + ENERGY\ (ATP)$$

Thus when these two pathways are combined:

CARBON CYCLE:

LIGHT ➔ ENERGY (ATP)

And the derivative of *photosynthesis* is "made by the sun". And as we know, the Son of God gives us Light, and the Spirit gives us breath to bear fruits, and the Father sustains us with the water of life. Thus the overall pathway from photosynthesis becomes:

THE SON'S FOLLOWERS:

SPIRIT + SON + FATHER ➔ FRUITS + SPIRIT

And the derivative of *respiration* is "to breath". For the Spirit gives us the breath of life. Thus the overall pathway for cellular respiration becomes:

THE BREATH OF THE SPIRIT:

FRUITS + SPIRIT ➔ SPIRIT + FATHER + LIFE

And when these two pathways are combined:

THE CYCLE OF HIS FOLLOWERS:

SON ➔ LIFE

"For just as the Father raises the dead and gives them life, even so the Son also gives life to whom He wishes." (John 5:21)

But when man is buried in the ground, his flesh is degraded into oil. And oil burns fervently in the fire. For the pathway of the combustion of oil is:

HYDROCARBON COMBUSTION:

$$CH_4 + 2\,O_2 \rightarrow CO_2 + 2\,H_20 + \text{LIGHT} + \text{HEAT}$$

And the derivative of *combustion* is "to burn in the fire". For the chaff of mankind is like oil, and their bodies shall burn in the fire. Thus the overall pathway of God's judgment becomes:

FIRE OF GOD'S JUDGMENT:

CHAFF + SPIRIT ➔ SPIRIT + FATHER + SON + *HEAT*

"Therefore thus says the Lord God, 'Behold, My anger and My wrath will be poured out on this place, on man and on beast and on the trees of the field and on the fruit of the ground; and it will burn and not be quenched.'" (Jeremiah 7:20)

The Carbon Cycle of Creation

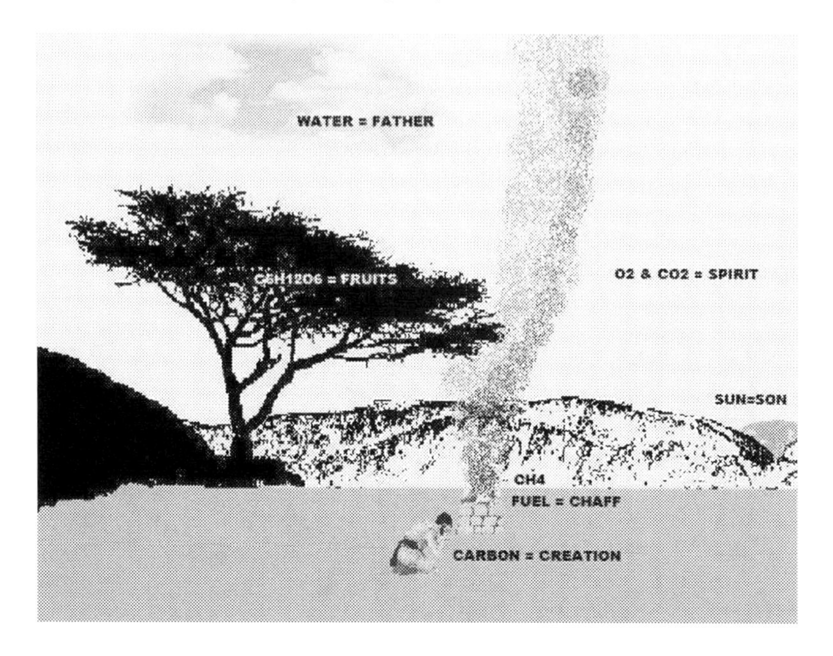

For dust is the carbon of this earth, and the flesh has the stigma of "six". For as God created the world in six days, He commanded His people to work in six days. (Exodus 20:11) And He commanded them to gather the wheat in six days. (Exodus 16:26) And for six years His people would yield from the land. (Exodus 23:10) And for six years the servants and livestock would labor. (Exodus 23:12) But after six years the servants were freed. (Exodus 21:2) And after six years all men shall be free. (Daniel 12)

And the tabernacle was built in six parts, to prepare for the Lamb that would come. For the lampstand of the tabernacle was made of six branches. (Exodus 25:32) And the wall of the tabernacle was held by six boards. (Exodus 26:22) And the ephod of the priests held a pair of six stones. (Exodus 28:10) And the table of the showbread held a pair of six loaves. (Leviticus 24:6)

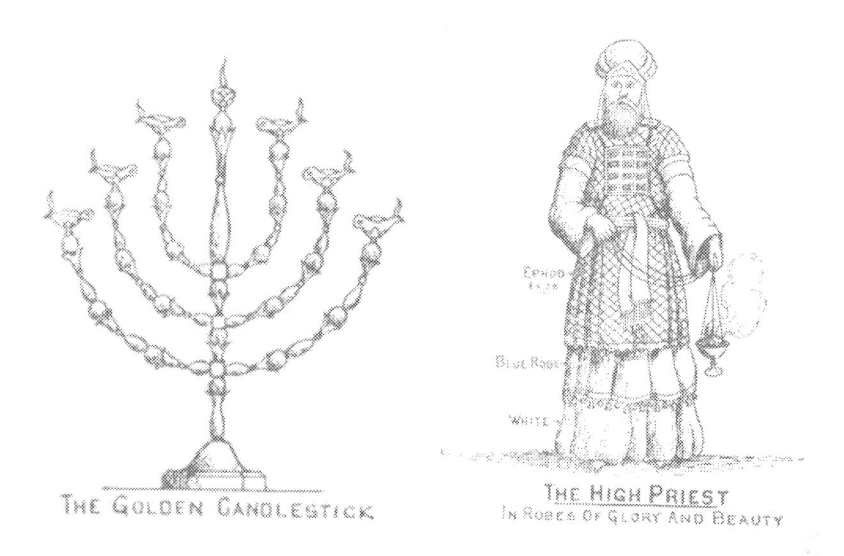

THE GOLDEN CANDLESTICK

THE HIGH PRIEST
IN ROBES OF GLORY AND BEAUTY

And the temple of God was constructed in six, to prepare for the King that would come. And the walls of the temple were closed for six days. (Ezekiel 46:1) And there were six stairs that led to the temple. (1 Kings 10:19) And on the six stairs that led to the throne, were six lions that guarded each side. (2 Chronicles 9:19) For the pair of six lions that guarded the throne, are the pair of six apostles that will guard the New City. "She had a great and high wall with twelve gates, and twelve angels at the gates, and names written on them, which are the names of the twelve tribes of the children of Israel: three gates on the east, three gates on the north, three gates on the south, and three gates on the west. Now the wall of the city had twelve foundations, and on them were the names of the twelve apostles of the Lamb." (Revelation 21:12-14)

And God taught His people to wait patiently in six, for "He shall deliver you in six troubles, yes, in seven no evil shall touch you." (Job 5:19) For His servant would wait on Mount Sinai for six days, before the Lord answered his prayers. (Exodus 24:16) And His people would march around Jericho six days, before its stone walls would fall down. (Joshua 6:3) And His apostles would wait for six days on the mountain before the angels appeared. (Mark 9:2) And the six days of waiting for the Sabbath was abolished, for the Son walked on earth that all men could be free. (Luke 13:4) For His servants no longer eat unleavened bread, but now eat the leaven of the Spirit. (Deuteronomy 16:8)

But woe to those who forsake Him, for the sixth seal shall be opened, and the sixth bowl shall be poured. For on January 3rd of 2010, the sixth seal shall be opened and New York City shall quake, and the stars of Wall Street shall fall from the heavens. (Revelations 6:12) And when the sixth bowl is poured out, the angels that guard the Euphrates River, will pour out their wrath with great fury. And what is the great Euphrates River? For the great Euphrates River was there, when the Garden of Eden was made, for "the fourth river is the Euphrates." (Genesis 2:14) And it was foretold that man's blood would flow free. "For the Lord God of hosts has a sacrifice. In the north country by the River Euphrates." (Jeremiah 46:10) For the river is fed by the Suez to the north, and its shallowest depth is just eight meters deep. And the river is fed by the Bab to the south, and its shallowest depth is just thirty meters deep. For the River Euphrates Depth (RED) shall cause the RED Sea to go dry. And the second woe is the great earthquake, the great quake of Mecca and Jerusalem. For the Dead Sea Rift and the Red Sea Fault, shall close off the Suez to the north, and the Bab el Mandab to the south. Then the sixth bowl shall be poured out. "When the sixth angel poured out his bowl on the great river Euphrates, and its water was dried up, so that the way of the kings from the east might be prepared." (Revelations 16:12)

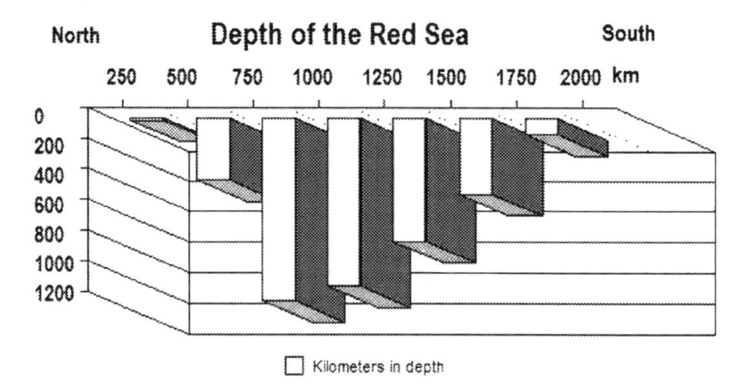

"'Therefore behold, the days are coming,' says the Lord, 'when it will no more be called Tophet, or the Valley of the Son of Hinnom, but the Valley of Slaughter; for they will bury in Tophet until there is no room.'" (Jeremiah 7:32)

The Sixth Seal and the Sixth Bowl of Wrath

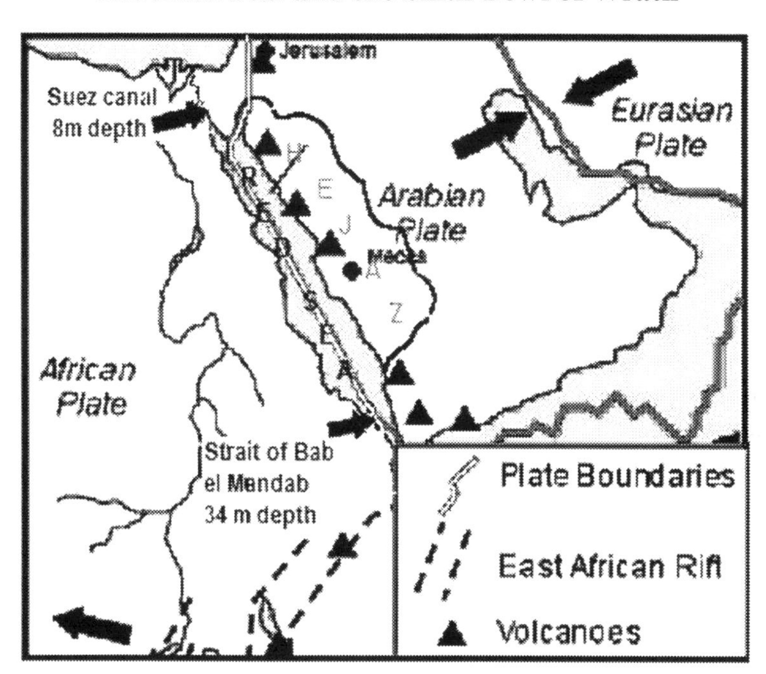

River Euphrates Depth: The RED Sea

"Order the buckler and shield, and draw near to battle! Harness the horses, and mount up, you horsemen! Stand forth with your helmets, polish the spears, put on the armor! Why have I seen them dismayed and turned back? Their mighty ones are beaten down; they have speedily fled, and did not look back, for fear was all around," says the Lord. "Do not let the swift flee away, nor the mighty man escape; they will stumble and fall toward the north, by the River Euphrates. "Who is this coming up like a flood, whose waters move like the rivers? Egypt rises up like a flood, and its waters move like the rivers; and he says, 'I will go up and cover the earth, I will destroy the city and its inhabitants.' Come up, O horses, and rage, O chariots! And let the mighty men come forth: the Ethiopians and the Libyans who handle the shield, and the Lydians who handle and bend the bow. For this is the day of the Lord God of hosts, a day of vengeance, that He may avenge Himself on His adversaries. The sword shall devour; it shall be satiated and made drunk with their blood; for the Lord God of hosts has a sacrifice in the north country by the River Euphrates. Go up to Gilead and take balm, O virgin, the daughter of Egypt; in vain you will use many medicines; you shall not be cured. The nations have heard of your shame, and your cry has filled the land; for the mighty man has stumbled against the mighty; they both have fallen together." (Jeremiah 46:3-12)

VII

Seven: The Number of the Lord (NKJV)

To Andrew the Prophet

Completed January 13, 2008

God delivers His creation in seven

"And on the **seventh** day God ended His work which He had done, and He rested on the **seventh** day from all His work which He had done. God blessed the **seventh** day and sanctified it, because in it He rested from all His work which God had created and made." Genesis 2:2-3

"You shall take with you **seven** each of every clean animal, a male and his female; two each of animals that are unclean, a male and his female; also **seven** each of birds of the air, male and female, to keep the species alive on the face of all the earth." Genesis 7:2-3

"Then the ark rested in the **seventh** month, the seventeenth day of the month, on the mountains of Ararat." Genesis 8:4

"The **seven** good cows are **seven** years, and the **seven** good heads are **seven** years; the dreams are one." Genesis 41:26

"If you buy a Hebrew servant, he shall serve six years; and in the **seventh** he shall go out free and pay nothing." Exodus 21:2

"And **seven**$_1$ priests shall bear **seven**$_2$ trumpets of rams' horns before the ark. But the **seventh**$_3$ day you shall march around the city seven times, and the priests shall blow the trumpets. But it came to pass on the **seventh**$_4$ day that they rose early, about the dawning of the day, and marched around the city **seven**$_5$ times in the same manner. On that day only they marched around the city **seven**$_6$ times. And the **seventh**$_7$ time it happened, when the priests blew the trumpets, that Joshua said to the people: "Shout, for the Lord has given you the city!" Joshua 6:4,15, 16

“He shall deliver you in six troubles, Yes, in **seven** no evil shall touch you.” Job 5:19

God heals mankind in seven

“But if the bright spot is white on the skin of his body, and does not appear to be deeper than the skin, and its hair has not turned white, then the priest shall isolate the one who has the sore **seven** days.” Leviticus 13:4

“The priest shall examine the plague and isolate that which has the plague **seven** days.” Leviticus 13:50

“If a woman has a discharge, and the discharge from her body is blood, she shall be set apart **seven** days; and whoever touches her shall be unclean until evening.” Leviticus 15:19

“And if any man lies with her at all, so that her impurity is on him, he shall be unclean **seven** days; and every bed on which he lies shall be unclean.” Leviticus 15:24

“He who touches the dead body of anyone shall be unclean **seven** days.” Numbers 19:11

God’s feasts were in seven

“On the fourteenth day of the first month at twilight is the Lord’s Passover. And on the fifteenth day of the same month is the Feast of Unleavened Bread to the Lord; **seven** days you must eat unleavened bread. On the first day you shall have a holy convocation; you shall do no customary work on it. But you shall offer an offering made by fire to the Lord for seven days. The **seventh** day shall be a holy convocation; you shall do no customary work on it.” Leviticus 23:5-8

“‘When you come into the land which I give to you, and reap its harvest, then you shall bring a sheaf of the firstfruits of your harvest to the priest. He shall wave the sheaf before the Lord, to be accepted on your behalf; on the day after the Sabbath the priest shall wave it... And you shall count for yourselves from the day after the Sabbath, from the day that you brought the sheaf of the wave offering: **seven** Sabbaths shall be completed." Leviticus 23:10-11,15

"Then the Lord spoke to Moses, saying, 'Speak to the children of Israel, saying: 'In the **seventh** month, on the first day of the month, you shall have a sabbath-rest, a memorial of blowing of trumpets, a holy convocation.''" Leviticus 23:23-24

"And the Lord spoke to Moses, saying: 'Also the tenth day of this **seventh** month shall be the Day of Atonement.''" Leviticus 23:26-27

"Speak to the children of Israel, saying: 'The fifteenth day of this **seventh** month shall be the Feast of Tabernacles for seven days to the Lord.''" Leviticus 23:34

"You shall dwell in booths for **seven** days. All who are native Israelites shall dwell in booths" Leviticus 23:42

"At the end of every **seven** years, at the time of the year of remission of debts, at the Feast of Booths, when all Israel comes to appear before the Lord your God at the place which He will choose, you shall read this law in front of all Israel in their hearing. 'Assemble the people, the men and the women and children and the alien who is in your town, so that they may hear and learn and fear the Lord your God, and be careful to observe all the words of this law. Their children, who have not known, will hear and learn to fear the Lord your God, as long as you live on the land which you are about to cross the Jordan to possess .''" Deuteronomy 31:10-13

"For **seven** days you shall offer an offering made by fire to the Lord. On the eighth day you shall have a holy convocation, and you shall offer an offering made by fire to the Lord. It is a sacred assembly, and you shall do no customary work on it." Leviticus 23:36

"'And you shall count **seven** sabbaths of years for yourself, **seven** times **seven** years; and the time of the **seven** sabbaths of years shall be to you forty-nine years." Leviticus 25:8

The Temple, the Sacrifice, and the priests were prepared in seven

"You shall make **seven** lamps for it, and they shall arrange its lamps so that they give light in front of it." Exodus 25:37

"The holy garments of Aaron shall be for his sons after him, that in them they may be anointed and ordained. For **seven** days the one of his sons who is priest in his stead shall put them on when he enters the tent of meeting to minister in the holy place. you shall do to Aaron and to his sons, according to all that I have commanded you; you shall ordain them through **seven** days. Each day you shall offer a bull as a sin offering for atonement, and you shall purify the altar when you make atonement for it, and you shall anoint it to consecrate it. For **seven** days you shall make atonement for the altar and consecrate it; then the altar shall be most holy, and whatever touches the altar shall be holy." Exodus 29:29-30, 35-37

"He sprinkled some of it on the altar **seven** times, anointed the altar and all its utensils, and the laver and its base, to consecrate them." Leviticus 8:11

"And you shall not go outside the door of the tabernacle of meeting for **seven** days, until the days of your consecration are ended. For **seven** days he shall consecrate you." Leviticus 8:33

"The priest shall dip his finger in the blood and sprinkle some of the blood **seven** times before the Lord, in front of the veil of the sanctuary." Leviticus 4:6

"Moreover the light of the moon will be as the light of the sun, And the light of the sun will be **sevenfold**, As the light of **seven** days, In the day that the Lord binds up the bruise of His people And heals the stroke of their wound." Isaiah 30:26

And the seven seals will be opened, and the seven bowls will be poured

"But one of the elders said to me, 'Do not weep. Behold, the Lion of the tribe of Judah, the Root of David, has prevailed to open the scroll and to loose its **seven** seals.'" Revelation 5:5

"When He opened the **seventh** seal, there was silence in heaven for about half an hour. And I saw the **seven** angels who stand before God, and to them were given **seven** trumpets." Revelation 8:2

"and cried with a loud voice, as when a lion roars. When he cried out, **seven** thunders uttered their voices." Revelation 10:3

"but in the days of the sounding of the **seventh** angel, when he is about to sound, the mystery of God would be finished, as He declared to His servants the prophets." Revelation 10:7

"In the same hour there was a great earthquake, and a tenth of the city fell. In the earthquake **seven** thousand people were killed, and the rest were afraid and gave glory to the God of heaven." Revelation 11:13

"Then the **seventh** angel sounded: And there were loud voices in heaven, saying, "The kingdoms of this world have become the kingdoms of our Lord and of His Christ, and He shall reign forever and ever!" Revelation 11:15

"Then I saw another sign in heaven, great and marvelous: seven angels having the **seven** last plagues, for in them the wrath of God is complete." Revelation 15:1

"Then the **seventh** angel poured out his bowl into the air, and a loud voice came out of the temple of heaven, from the throne, saying, 'It is done!'" Revelation 16:17

The Lord Shall Return in Seven

"Trust in the Lord with all your heart, and lean not on your own understanding; in all your ways acknowledge Him, and He shall direct your paths." (Proverbs 3:5-6) Who can understand the Lord, the almighty One who creates and destroys? For the Lord has power and dominion over us all. But in His infinite mercy and grace, He has granted mankind His favor, for He shall forgive all men in seven, for in "two thousand three hundred days; then the sanctuary shall be cleansed." (Daniel 8:14)

For all things were perfected in seven. For the Lord delivered mankind in seven. For on the seventh day, the Lord would rest from creation, so that man could prosper and live. (Genesis 2:2-3) And when God removed all evil from the earth, He commanded that Noah place in the arc, seven of each creature that they may live. (Genesis 7:2-3) And it was in the seventh month of that year, that God saved His creation from the flood. (Genesis 8:4) And when God sent Joseph to Egypt, the pharaoh stored up produce for seven years, that all of the nations could live. (Genesis 41:26) And the servants were released in seven years, so that they could live and be free. (Exodus 21:2) And Joshua camped by Jericho for seven$_1$ days, and on the seventh$_2$ day marched around the city seven$_3$ times, and the seven$_4$ priests blew their seven$_5$ trumpets seven$_6$ times. "And the ***seventh***$_7$ time it happened, when the priests blew the trumpets, that Joshua said to the people: 'Shout, for the Lord has given you the city!'" (Joshua 6:4,15,16) For the Lord has promised we shall live in ***seven***, for "He shall deliver you in six troubles, yes, in ***seven*** no evil shall touch you." (Job 5:19)

And the Lord healed His people in seven days. For anyone with a sore was isolated for seven days. (Leviticus 13:4) And anyone with the plague was isolated for seven days. (Leviticus 13:50) And any woman with a discharge was isolated for seven days. (Leviticus 15:19) And any man with her was isolated for seven days. (Leviticus 15:24) And anyone who touched a dead body was isolated for seven days. (Numbers 19:11) But soon there will be no isolation, for the dead bodies shall litter the streets.

And in remembrance of His faithfulness, the feasts of the Lord are in seven. And at the Feast of the Passover (Pesach) and Unleavened Bread (Hag Hamatzot), for seven days they ate the unleavened bread. (Leviticus 23:5-8) And they gave up an offering at the Feast of Firstfruits. (Yom HaBikkurim) And an offering at the seventh Sabbath at the Feast of the Weeks. (Shavnot) (Leviticus 23:10-11,15) And in remembrance of their passage through the wilderness, as an offering at the Feast of the Booths (Sukkot), they would dwell in the booths for seven days. (Leviticus 23:34,42) And every seventh year on the Feast of the Booths, His people read aloud the words of the Torah. (Simchat Torah) (Deuteronomy 31:10-13) And after the Torah was read seven times, on the seventh month on the Day of Atonement, a Jubilee of His faithfulness was celebrated. (Leviticus 25:8)

And the temple was built in seven for the Lamb. And seven lampstands sat in front of the tabernacle. And in seven days the Passover was prepared by the priests. (Exodus 29:29-30,35-37) And for seven days the priests remained in the tent. (Leviticus 8:33) And seven times the priests were consecrated by oil. (Leviticus 8:11) And seven times the blood was spread on the veil. (Leviticus 4:6) And the prophets foretold, when the sacrifice was complete, that the sun would shine sevenfold. (Isaiah 30:26)

And in seven the sacrifice of the Lamb was prepared. For His path was prepared by one of seven lampstands. (Matthew 3:1) And seven days before the sacrifice of the Lamb, the path was prepared to enter the city. "When they heard that Jesus was coming to Jerusalem, they took branches of palm trees and went out to meet Him, and cried out: 'Hosanna! Blessed is He who comes in the name of the Lord! The King of Israel!'" (John 12:12-13) And the Lamb was anointed with oil, from the woman who was delivered from seven. "For in pouring this fragrant oil on My body, she did it for My burial." (Matthew 26:12/Mark 16:9) And when He was sacrificed on the cross, He cried out seven words and the veil tore in two. "Then, behold, the veil of the temple was torn in two from top to bottom; and the earth quaked, and the rocks were split." (Matthew 27:51) And there were seven words that he cried: "Eloi, Eloi, why hast thou forsaken me?" (Mark 15:34) For He cried out seven words that Satan would be deceived, to believe that the Father had abandoned His Son. Thus the Son was removed from the tomb, so that Satan might bind Him in hell. But to Satan's dismay, it was the Father who delivered His Son down to hell, that the prisoners of hell could be free. For the Son set them free and bound Satan who is the Devil that serpent of old in hell. "He laid hold of the dragon, that serpent of old, who is the Devil and Satan, and bound him for a thousand years." (Revelation 20:2) Then the Son ascended into heaven and opened the gates of the heavens, and the Son shone forth sevenfold, for it is He who "binds up the bruise of His people and heals the stroke of their wound." (Isaiah 30:26) And the time is now upon us, for the seven seals soon will be opened. And in the year of two thousand and seven, the first seal was broken for Sadr the antichrist had arrived. (Revelation 5:5) And the seven lampstands will blow seven trumpets, to warn of the seven bowls of wrath that will come. (Revelation 8:2) And after the seven lampstands are martyred, the seventh seal shall be broken. (Revelation 10:3,7) And when the seven angels have arisen, Jerusalem shall fall and seven thousand shall perish. (Revelation 11:13) And the seven angels shall shout aloud from the temple, "the kingdoms of this world have become the kingdoms of our Lord and of His Christ, and He shall reign forever and ever!" (Revelation 11:15) And the seven angels shall be handed the seven bowls of wrath. "The seven angels having the seven last plagues, for in them the wrath of God is complete." (Revelation 15:1) And when the seventh bowl is poured out, the seventh angel shall proclaim, "It is done!" (Revelation 16:17) "Then a mighty angel took up a stone like a great millstone and threw it into the sea, saying, 'Thus with violence the great city Babylon shall be thrown down, and shall not be found anymore. The sound of harpists, musicians, flutists, and trumpeters shall not be heard in you anymore. No craftsman of any craft shall be found in you anymore, and the sound of a millstone shall not be heard in you anymore. The light of a lamp shall not shine in you anymore, and the voice of bridegroom and bride shall not be heard in you anymore. For your merchants were the great men of the earth, for by your sorcery all the nations were deceived. And in her was found the blood of prophets and saints, and of all who were slain on the earth." (Revelations 18:21-24)

VIII

Everything is Perfected in Eight (NASB)

To Andrew the Prophet

Completed January 15, 2008

The foreskin of the flesh is removed in eight

"And every male among you who is **eight** days old shall be circumcised throughout your generations, a servant who is born in the house or who is bought with money from any foreigner, who is not of your descendants." Genesis 17:12

"Then Abraham circumcised his son Isaac when he was **eight** days old, as God had commanded him." Genesis 21:4

"And He gave him the covenant of circumcision; and so Abraham became the father of Isaac, and circumcised him on the **eighth** day; and Isaac became the father of Jacob, and Jacob of the twelve patriarchs." Acts 7:8

"On the **eighth** day the flesh of his foreskin shall be circumcised." Leviticus 12:3

"And when **eight** days had passed, before His circumcision, His name was then called Jesus, the name given by the angel before He was conceived in the womb." Luke 2:21

The outer court of the temple is cleansed in eight

"Now they began the consecration on the first day of the first month, and on the **eighth** day of the month they entered the porch of the Lord. Then they consecrated the house of the Lord in **eight** days, and finished on the sixteenth day of the first month." 2 Chronicles 29:17

"Its porches were toward the outer court; and palm tree ornaments were on its side pillars, and its stairway had **eight** steps." Ezekiel 40:31

"Four tables were on each side next to the gate; or, **eight** tables on which they slaughter sacrifices." Ezekiel 40:41

The Final Temple

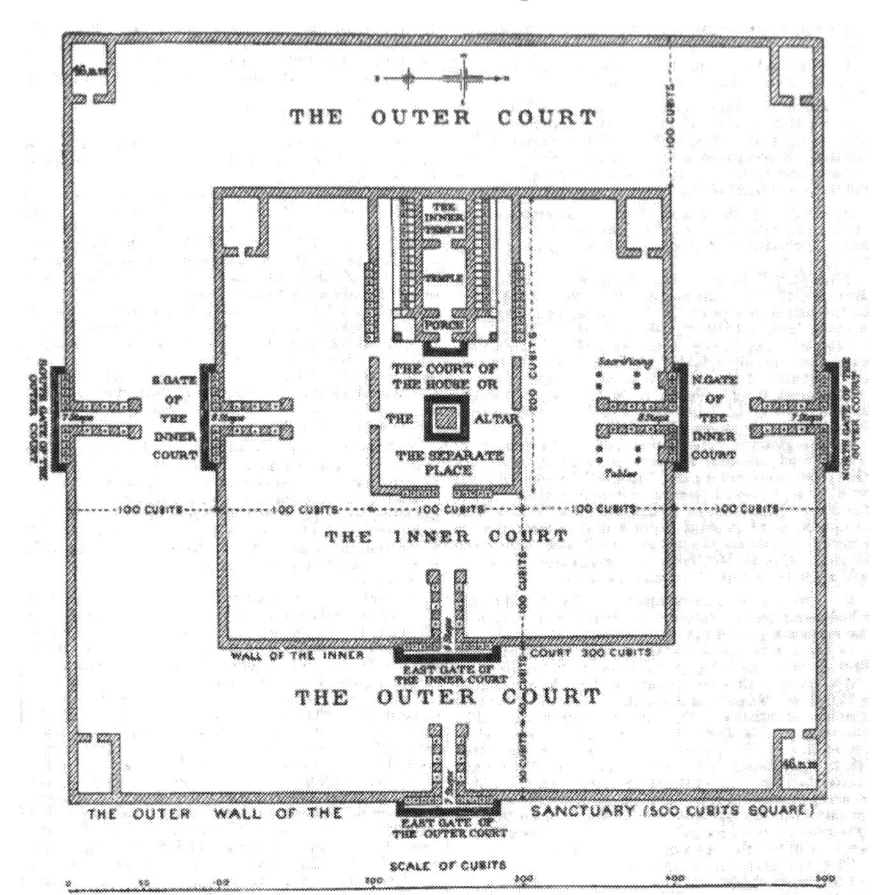

God reconciles all in eight

"who once were disobedient, when the patience of God kept waiting in the days of Noah, during the construction of the ark, in which a few, that is, **eight** persons, were brought safely through the water." 1 Peter 3:20

"Then the **eighth** day he shall bring them for his cleansing to the priest, at the doorway of the tent of meeting, before the Lord." Leviticus 14:23

"Then on the **eighth** day he shall take for himself two turtledoves or two young pigeons, and come before the Lord to the doorway of the tent of meeting and give them to the priest" Leviticus 15:14

"For seven days you shall present an offering by fire to the Lord. On the **eighth** day you shall have a holy convocation and present an offering by fire to the Lord; it is an assembly. You shall do no laborious work." Leviticus 23:36

"When you are sowing the **eighth** year, you can still eat old things from the crop, eating the old until the ninth year when its crop comes in." Leviticus 25:22

And the eighth ruler will be destroyed

"When they have completed the days, it shall be that on the **eighth** day and onward, the priests shall offer your burnt offerings on the altar, and your peace offerings; and I will accept you,' declares the Lord God." Ezekiel 43:27

"The beast which was and is not, is himself also an **eighth** and is one of the seven, and he goes to destruction." Revelations 17:11

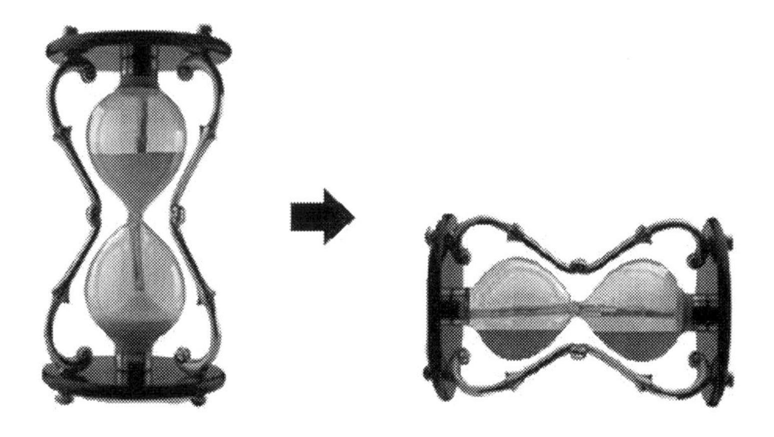

When the 8th Ruler is Defeated, God's Kingdom Will Be ∞

We all have heard of the term seventh heaven, but have you not heard of the eighth? For the prophecy speaks of the eighth, "the beast which was and is not, is himself also an ***eighth*** and is one of the seven, and he goes to destruction." (Revelations 17:11) For the temple of God will be perfected in seven, and the rest of mankind will be perfected in eight. And when the 8th ruler is destroyed, God's kingdom shall be infinite forever (∞). For the ordinance of time is the seven-fold week. But the eighth ruler who is "time" shall be slain, and death shall no longer hold man.

Eight is the atomic number of oxygen, and as we know oxygen is the breath of life. For as oxygen is the breath of life, so is the Spirit the breath of man's life. For "the Spirit of God has made me, and the breath of the Almighty gives me life." (Job 33:4) And the diatomic structure of oxygen is an eight. But when the end comes, His breath shall depart, "for I will not contend forever, nor will I always be angry; for the spirit would grow faint before Me, and the breath of those whom I have made." (Isaiah 57:16)

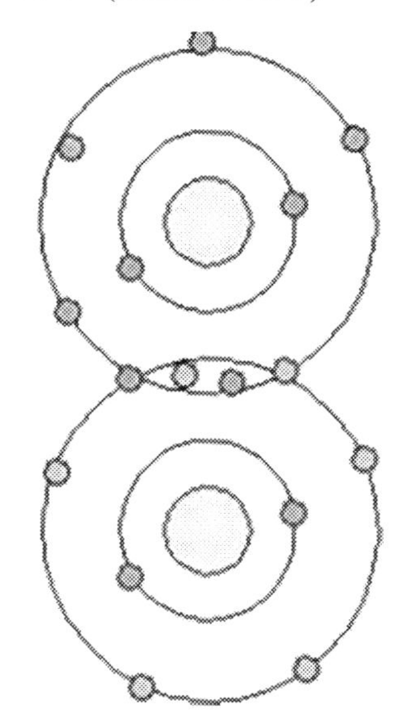

The atomic structure of oxygen

Eight is derived from the Indian word *yettu* that is in the shape of an **H**. And the number of eight is the Greek letter *eta* that is in the shape **H**. And in science **H** represents enthalpy or heat. And at the end "the day of the Lord will come like a thief, in which the heavens will pass away with a roar and the elements will be destroyed with intense **heat**, and the earth and its works will be burned up." (2 Peter 3:10)

Enthalpy is the thermodynamic potential of a system. And this is the sum of the system's internal energy and its work. And work is equivalent to pressure times volume. Thus the formula for enthalpy is:

$$\text{Enthalpy} = \text{Internal Energy} + \text{Work}$$

$$H = U + PV$$

H represents enthalpy
U represents internal energy
P represents pressure
V represents volume

And the change in the potential of a system, is equivalent to the change of internal energy, plus the work that is done in the system ($p\Delta V$). Thus:

$$\Delta H = \Delta U + p\Delta V$$

And as it says "if anyone is in Christ, he is a new creature; the old things passed away; behold, new things have come." (2 Corinthians 5:17) And it also says "you see that a man is justified by works and not by faith alone." (James 2:24)

Thus:

New man in Christ (ΔU)

\+ **Man's good works** ($p\Delta V$)

Man's just rewards after the fire (ΔH)

For "each man's work will become evident; for the day will show it because it is to be revealed with fire, and the fire itself will test the quality of each man's work. If any man's work which he has built on it remains, he will receive a reward. If any man's work is burned up, he will suffer loss; but he himself will be saved, yet so as through fire." (1 Corinthians 3:13-15)

And man's work began from the days of our forefathers, for the foreskin of man's flesh was removed on the eighth. And it was on the eighth day, that Abraham and his sons and his servants were circumcised. (Genesis 17:12) For Isaac was circumcised on the eighth day. (Genesis 21:4) And the forefathers were circumcised on the eighth day. (Acts 7:8) And each male child was circumcised on the eighth day. (Leviticus 12:3) And our Lord was circumcised on the eighth day. (Luke 2:21) For a new man is formed by the circumcision of the flesh. For circumcision comes not from the foreskin of our loins, but from the foreskin of enslavement to this world. But "'behold, the days are coming,' declares the Lord, 'that I will punish all who are circumcised and yet uncircumcised.'" (Jeremiah 9:25)

And the court of the second temple was built in eight, in semblance of the judgment to come. For when the temple was restored, for eight days the courtyard was cleansed. (2 Chronicles 29:17) And the new temple shall be built in eight, in semblance of the judgment to come. For there will be eight steps that lead to the court. (Ezekiel 40:31) And there will be eight tables of sacrifice in the court. (Ezekiel 40:41) And the prophecy foretold, that the court outside the temple would be cleansed. "Leave out the court which is outside the temple and do not measure it, for it has been given to the nations; and they will tread under foot the holy city for forty-two months." (Revelation 11:2) For each man in the court, must be sacrificed in the court, before he may enter the temple of God.

But after the judgment, God shall save mankind in eight. For after the wrath of the flood, eight people were saved in the ark. (1 Peter

3:20) And after eight days the lepers were cleansed, that they could enter the tent. (Leviticus 14:23) And any man whose skin had a sore, after eight days of cleansing could enter the tent. (Leviticus 15:14) And on the eighth day, the Day of Atonement was held. (Leviticus 23:36) And on the eighth year after seven times seven years, the Jubilee of His people was celebrated. (Leviticus 25:22)

And the prophets foretold that this day would come. And the fire of judgment shall burn, before man shall be accepted by God. For "'when they have completed the days, it shall be that on the eighth day and onward, the priests shall offer your burnt offerings on the altar, and your peace offerings; and I will accept you,' declares the Lord God." (Ezekiel 43:27) And after the seventh bowl is poured out, "the beast which was and is not, is himself also an eighth and is one of the seven, and he goes to destruction." (Revelations 17:11) And the great riddle is who are the seven, and who is the eighth that goes to destruction? For the scripture reveals the answer to this question. For the seven rulers are the seven angels of Satan, who rule the seven realms which exist. And their names represent the seven-fold nature of the week. And "the first is Athoth...the second is Eloaios...the third is Astaphaios...the fourth is Yao...the fifth is Sabaoth...the sixth is Adonin..and the seventh is Sabbataios...this is the sevenfold nature of the week". (Apocrypha of John) For it is time which is the beast and the last enemy of mankind. And the eighth beast who "was and is not" is Sabaoth, for "seven archangels stand before the throne. Sabaoth is the eighth and he has authority". (Nag Hammadi "*On the Origin of the World*").

And when the eighth ruler is destroyed, the kingdom of God shall be complete, and the Lord shall reign forever and ever. And "the Lamb will overcome them, because He is Lord of lords and King of kings, and those who are with Him are the called and chosen and faithful." (Revelation 17:14) "And there will no longer be any night; and they will not have need of the light of a lamp nor the light of the sun, because the Lord God will illumine them; and they will reign forever and ever." (Revelation 22:5)

Pieta **Michelangelo**

PIETA

Pi + Eta = π + H = 80 + 8

88

IX

Six to Nine: Darkness Shall Consume the Earth (NASB)

To Andrew the Prophet

Completed January 18, 2008

The year of Jubilee is completed in the ninth

"When you are sowing the eighth year, you can still eat old things from the crop, eating the old until the **ninth** year when its crop comes in." Leviticus 25:22

The kingdom is attacked in the ninth

"In the **ninth** year of Hoshea, the king of Assyria captured Samaria and carried Israel away into exile to Assyria, and settled them in Halah and Habor, on the river of Gozan, and in the cities of the Medes." 2 Kings 17:6

"Now in the **ninth** year of his reign, on the tenth day of the tenth month, Nebuchadnezzar king of Babylon came, he and all his army, against Jerusalem, camped against it and built a siege wall all around it ." 2 Kings 25:1

Darkness consumed the land until the ninth hour

"When the sixth hour (noon) came, darkness fell over the whole land until the **ninth** hour. (3 pm)" Mark 15:33

Christ the Lamb was consumed in the ninth hour

"At the **ninth** hour (3 pm) Jesus cried out with a loud voice, 'Eloi, Eloi, lama sabachthani' which is translated, 'My God, My God, why have You forsaken Me?' When some of the bystanders heard it, they began saying, 'Behold, He is calling for Elijah.' Someone ran and filled a sponge with sour wine, put it on a reed, and gave Him a drink, saying, 'Let us see whether Elijah will come to take Him down.' And Jesus uttered a loud cry, and breathed His last." Mark 15:34-37

And the Lord shall return for the remaining nine

"Were there not ten cleansed? But the **nine**—where are they? 'Was no one found who returned to give glory to God, except this foreigner?'" Luke 17:17-18

The Ninth Hour and the Black Hole

We all wait in trepidation for the ninth hour to come. For Israel was defeated in the ninth year of Hoshea, for they "had sinned against the Lord their God." (2 Kings 17:6-7) And in the ninth year of Nebuchadnezzar's rule, he overtook the city and "burned the house of the Lord". (2 Kings 25:1-9) And the nine days of Av are the "The Nine Days" mourning (Tisha HaYamim), in remembrance of the temple that was burned. And in the ninth year after the Jubilee, His people waited anxiously for the harvest to return. (Leviticus 25:22)

For "when the sixth hour came, darkness fell over the whole land until the ninth hour." (Mark 15:33) For when the Son was nailed to the cross, darkness fell over all the land. But the darkness fell not from the moon, for a solar eclipse lasts only three minutes, but darkness would fall for three hours. And the darkness would cover the whole land, for the Son was nailed to the cross. And darkness fell over all the land, for the Son and the Light had expired. And when the sixth seal is broken, the darkness shall fall once again. "I looked when He broke the sixth seal, and there was a great earthquake; and the sun became black as sackcloth made of hair, and the whole moon became like blood." (Revelation 6:12) And then men will know that end will come soon, for who can withstand the judgment indeed!

The symbol for nine is the Greek letter theta Θ. And theta is derived from the Phoenician letter *Teth,* which was written as a circle and a cross. And this cross with the sun is the symbol for the "earth". For the Son carried His cross on the earth. For "He humbled Himself by becoming obedient to the point of death, even death on a cross." (Philippians 2:8) And *theta* was written as a dot within a circle. And this ancient symbol represented the sun god in the clouds. And now the Son of Man sits and waits in the clouds. "Then I looked, and behold, a white cloud, and sitting on the cloud was one like a son of man." (Revelation 14:14)

And the value of Jesus Christ is 9, for the value of His name "Iesous Christos" is IX. And the derivative of theta is *theou* which means GOD. For God's name is "Iesous Christos **Theou** Yios Soter" (I X Θ Y ∑), for His name is "Jesus Christ, Son of God, Savior". And *theou* is derived from the sun god Helios, who drives with his chariot through the wind. "For behold, the Lord will come with fire and with His chariots, like a whirlwind, to render His anger with fury, and His rebuke with flames of fire. For by fire and by His sword the Lord will judge all flesh; and the slain of the Lord shall be many." (Isaiah 66:15-16)

And in Athens theta means "thanatos" or death. For this symbol was used as the penalty of death, when judging the fate of a prisoner. "And though they found no cause for death in Him, they asked Pilate that He should be put to death." (Acts 13:28) And when the 6th and 9th join together, the earth shall turn dark and the universe shall tremble. "Therefore I will make the heavens tremble, and the earth will be shaken from its place at the fury of the Lord of hosts In the day of His burning anger." (Isaiah 13:13)

6 + 9 = 69

For this symbol represents the dual nature of God and His creation. For this symbol represents yin and yang, the dual nature of man and creation. And this symbol represents the structure of DNA, the genome from Genesis that sustains all generations. And this symbol represents the dual nature of atoms, for the components of protons and neutrinos, are unified by spiral helical fields. (Positron and Neutrino Helix Spiral Fields /*Smart's Unified Helix Spiral Field Theory*) And this symbolizes the dual nature of the sun, and the magnetic field of their two poles. And this symbolizes the dual nature of the Son, and the dual nature of the Lion and the Lamb, and His dual nature to save and destroy. For when the 6th and the 9th come together, the symbol is the black hole of space and of time! And through the death of spacetime, the space of death shall be destroyed, for "the last enemy that will be destroyed is death". (1 Corinthians 15:26)

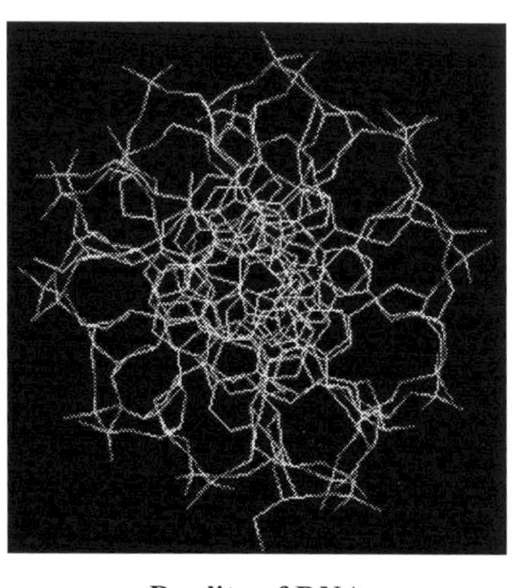

Duality of DNA

Duality of Man's Nature

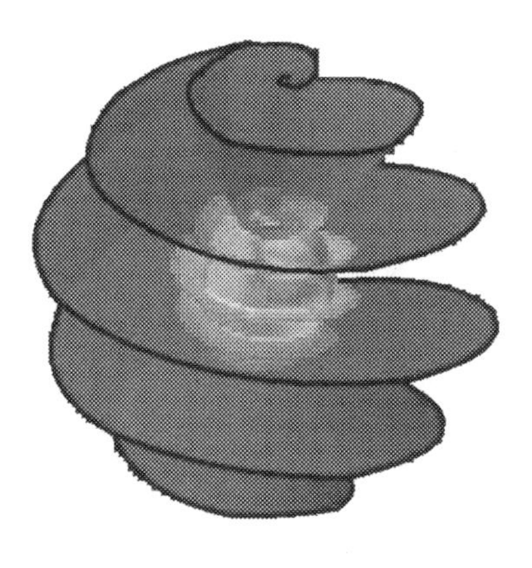

Duality of Subatomic Particles

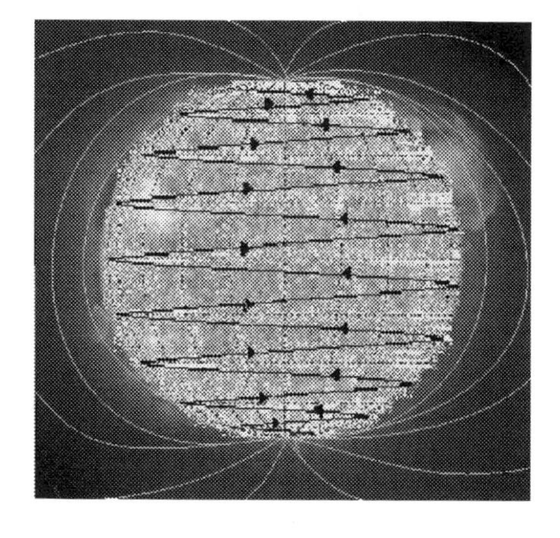

Duality of the Stars and the Planets

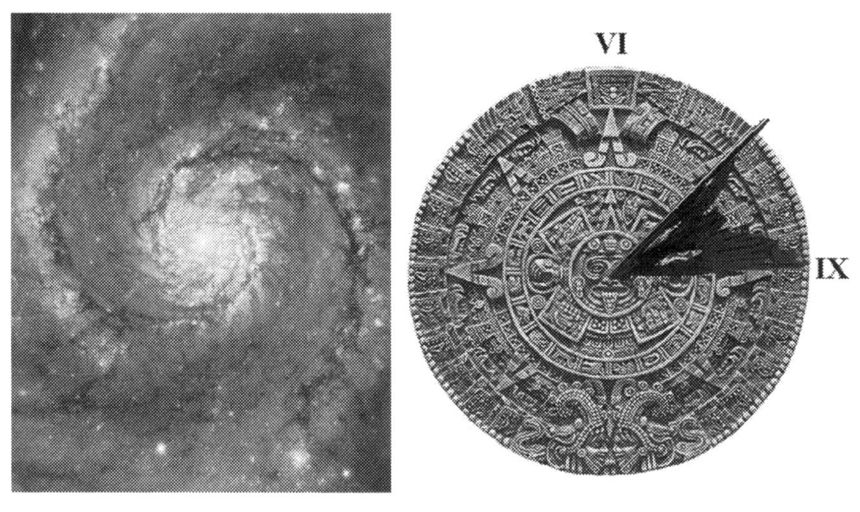

Duality of the Universe **Duality of Spacetime**

For scientist are correct in postulating, that the universe was created by a "Big Bang". For "in the beginning God created the heavens and the earth. The earth was without form, and void; and darkness was on the face of the deep." (Genesis 1:1-2) And that " B i g Bang" most scientist concur is a black hole.

A black hole is a region of space, in which the gravitational force is so strong, that not even light can escape from its grip. And surrounding the black hole is a region called the ergosphere, which entraps any mass within its vicinity. And when a mass is dragged in by its pull, that mass is accelerated to exceptional velocities, and eventually approaches the speed of light. And when velocity increases and approaches the speed of light, then *proper time* decreases and thus time slows down. And "who then is the faithful and sensible slave whom his master put in charge of his household to give them their food at the ***proper time***?" (Matthew 24:45) For thousands of black holes exist in our galaxy, but greater than ninety percent have yet to be found. And truly, scientists have recently concluded, that the center of our universe is a black hole. And its mass is billions of times greater than our solar system, and its width is many millions of light years wide.

Yet another concerning fact is that mass is decreasing. For the universal prototype is slowly losing mass. "The 118-year-old cylinder that is the international prototype for the metric mass, kept tightly under lock and key outside Paris, is mysteriously losing weight - if ever so slightly." (AP - September 12, 2007) (*In 2007, the universal 1 kg measured 0.999050 kg*)

And if we apply the laws of relativity (1) and the first law of thermodynamics (2):

(1) $$E = mc^2$$

(2) Energy is neither created nor destroyed.

Then: $$(1.0\ \text{kg})\ c_1^2 = (0.999050\ \text{kg})\ c_2^2$$

where c_1 is the original speed of light

and c_2 is the present speed of light

$$c_2 = \sqrt{0.9905}\ c_1$$

$$c_2 = 1.0005\ c_1$$

Then, the only logical explanation is that the speed of light is increasing.

And this would explain why global warming is occurring. For global warming is not due to carbon emissions, but is due to an increase in the speed of light. For when the sun delivers more photons to the ionosphere, the solar heating of our atmosphere thus increases. For the ionosphere is an ionized layer above the earth's atmosphere, which is dominated by ions (NO+, O2+ and O+) which are called "plasma". This plasma is affected by photoionization from UV rays. And when the energy from the sun increases, the plasma density increases in an exponential manner. And this increase in the ionosphere has a profound effect on the atmosphere below. For this shift in plasma density dynamically and chemically drives the solar currents, and alters the thermospheric wind patterns, and enhances the equatorial electrojet currents below. And the Mayans foretold that global warming would occur, for as they said "the solar winds would increase during the transition period of the final cycle (1999-2012)." (*Mayan calendar*)

And as we know, the symbol ***c*** represents the velocity of light, and velocity is defined as distance over time. Thus if the speed of light is increasing, then time must be decreasing. And when any mass like our galaxy, enters a black hole then ***time must slow down***. For when a mass enters a black hole, its proper time slows down through a process called *gravitational time dilation*. And as the Lord said, "when I choose the ***proper time***, I will judge uprightly. The earth and all its inhabitants are dissolved; I set up its pillars firmly." (Psalms 75:2-3)

And the center of a black hole is called the *event horizon*. And what happens when we pass through the *event horizon*? Physicists postulate that there are three phenomena which occur. The first is that temperature approaches absolute zero, and that entropy or disorder becomes zero. And as we know Satan holds the power of entropy, but his power shall be abolished at the end. (Addendum 3 from *Signs, Science, and Symbols of the Prophecy*) "For He must reign until He has put all His enemies under His feet. The last enemy that will be abolished is death." (1 Corinthians 15:20-26) The second phenomenon that occurs when we pass through the event horizon, is that we transition into a parallel universe. And physicists define these transition points as "wormholes". And after the seventh seal is opened, the wormhole shall be opened, and "the name of the star is called Wormwood". (Revelation 8:11) And the third phenomenon that occurs when we pass through the event horizon, is that we enter a "sea" of energy called zero-point energy. Physicists call this "sea" of energy, an infinitely dense region defined as "singularity". And singularity is the limit at which physical properties approach infinite values, and thus the laws of physics seize to exist. And at the singularity, matter and antimatter shall combine, and mass and energy shall become one. And once all is reconciled, a new energy and a new mass shall be revealed. And "then I saw a new heaven and a new earth; for the first heaven and the first earth passed away, and there is no longer any sea." (Revelation 21:1)

X

God's Hands Shall Redeem All (NASB)

To Andrew the Prophet

Completed January 23, 2008

God's covenants comes in ten

"So he was there with the Lord forty days and forty nights; he did not eat bread or drink water. And he wrote on the tablets the words of the covenant, the **Ten** Commandments." Exodus 34:28

God redeems mankind in ten

"The water decreased steadily until the **tenth** month; in the **tenth** month, on the first day of the month, the tops of the mountains became visible." Genesis 8:5

"Speak to all the congregation of Israel, saying, 'On the **tenth** of this month they are each one to take a lamb for themselves, according to their fathers' households, a lamb for each household." Exodus 12:3

"On exactly the **tenth** day of this seventh month is the day of atonement; it shall be a holy convocation for you, and you shall humble your souls and present an offering by fire to the Lord." Leviticus 23:37

God's Temple is held in His hands

"Moreover you shall make the tabernacle with **ten** curtains of fine twisted linen and blue and purple and scarlet material; you shall make them with cherubim, the work of a skillful workman." Exodus 26:1

"For the width of the court on the west side shall be hangings of fifty cubits with their **ten** pillars and their ten sockets." Exodus 27:12

"Also in the inner sanctuary he made two cherubim of olive wood, each **ten** cubits high. Five cubits was the one wing of the cherub and five cubits the other wing of the cherub; from the end of one wing to the end of the other wing were **ten** cubits. The other cherub was **ten** cubits; both the cherubim were of the same measure and the same form." 1 Kings 6:23-25

"Then he made the **ten** stands of bronze; the length of each stand was four cubits and its width four cubits and its height three cubits." 1 Kings 7:27

"He made **ten** basins of bronze, one basin held forty baths; each basin was four cubits, and on each of the **ten** stands was one basin." 1 Kings 7:38

"Now the leaders of the people lived in Jerusalem, but the rest of the people cast lots to bring one out of **ten** to live in Jerusalem, the holy city, while nine-tenths remained in the other cities." Nehemiah 11:1

But only one of ten returned to give glory to God

"Surely all the men who have seen My glory and My signs which I performed in Egypt and in the wilderness, yet have put Me to the test these **ten** times and have not listened to My voice" Number 14:22

"While He was on the way to Jerusalem, He was passing between Samaria and Galilee. As He entered a village, **ten** leprous men who stood at a distance met Him; and they raised their voices, saying, 'Jesus, Master, have mercy on us!' When He saw them, He said to them, 'Go and show yourselves to the priests.' And as they were going, they were cleansed. Now one of them, when he saw that he had been healed, turned back, glorifying God with a loud voice, and he fell on his face at His feet, giving thanks to Him. And he was a Samaritan. Then Jesus answered and said, 'Were there not **ten** cleansed? But the nine—where are they? Was no one found who returned to give glory to God, except this foreigner?' And He said to him, 'Stand up and go; your faith has made you well.'" Luke 17:11-19

"Thus says the Lord of hosts, 'In those days **ten** men from all the nations will grasp the garment of a Jew, saying, 'Let us go with you, for we have heard that God is with you.''" Zechariah 8:23

The Hands of God are the Hands of Justice

"For we know Him who said, 'Vengeance is Mine, I will repay.' And again, 'The Lord will judge His people.' It is a terrifying thing to fall into the hands of the living God.'" (Hebrews 10:30-31) For when the hands of the Son come together, the fire and the judgment shall come. For when the Son came as the Lamb, He said "all things have been handed over to Me by My Father, and no one knows who the Son is except the Father, and who the Father is except the Son, and anyone to whom the Son wills to reveal Him." (Luke 10:22) For when the Lamb came to earth, all things were handed to Him from the Father. And the Shepherd came for His people, for "My sheep hear My voice, and I know them, and they follow Me; and I give eternal life to them, and they will never perish; and no one will snatch them out of My hand." (John 10:27-28) But most men are not of His fold, and the hands of the Son shall return. But the Son shall return as the Lion, "and I will set My glory among the nations; and all the nations will see My judgment which I have executed and My hand which I have laid on them." (Ezekiel 39:21)

For ten represents the judgment of His hands. And the Roman numeral for ten is **X,** which is formed by the union of two **V**s. For as we know **V** is the right hand of God. For the Son sits on the right hand of the Father, and all things were handed from the Father to the Son. For the Son and the Priest, was anointed on the thumb of His right hand. "Moses slaughtered it and took some of its blood and put it on the lobe of Aaron's right ear, and on the thumb of his right hand." (Leviticus 8:23)

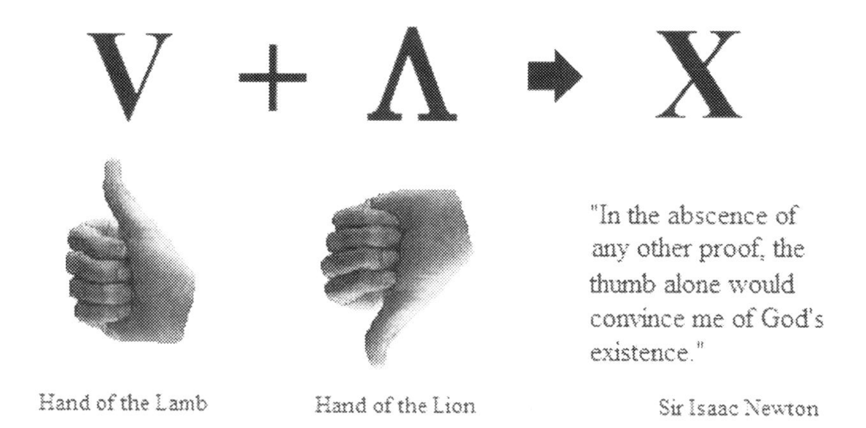

And the Greek numeral for ten is iota (ι). And as we know, iota is the power of the Spirit. (Addendum 4: Iota the Number of the Holy Spirit). For "on the other hand I am filled with power—with the Spirit of the Lord". (Micah 3:8) And the Roman symbol for ten is a cross (X). And the Asian symbol for ten is a cross.

For when man crosses God covenants must be made, and the covenants with mankind were in ten. For ten generations passed between Adam and Noah, when God destroyed man in the flood. And God made a covenant with Noah, not to destroy mankind with water again. (Genesis 5:29) And ten generations passed between Noah and Abram. And God made a covenant with Abraham, that he would father the nations. (Genesis 11:27) And the covenant with Israel was written on stone, by the hands of the Lord on Mount Sinai. (Exodus 34:28) And the Lamb made a new covenant with His people, for "He bore our sins in His body on the cross". (1 Peter 2:24) And "for this reason He is the mediator of a new covenant, so that, since a death has taken place for the redemption of the transgressions that were committed under the first covenant, those who have been called may receive the promise of the eternal inheritance." (Hebrews 9:15)

And in the year of 2010, the Son shall return to seal the last covenant. But woe to those who are left behind, "for a covenant is valid only when men are dead." (Hebrews 9:17) And when the seventh seal is opened and the seventh trumpet is blown, the final covenant shall be sealed with the blood of mankind. "And the temple of God which is in heaven was opened; and the ark of His covenant appeared in His temple, and there were flashes of lightning and sounds and peals of thunder and an earthquake and a great hailstorm." (Revelation 11:19)

But God in His mercy redeems all men in ten. For during the flood, it was "in the tenth month, on the first day of the month, the tops of the mountains became visible." (Genesis 8:5) And it was on the tenth day of the month of the Passover, that His people would sacrifice the lamb. (Exodus 12:3) And it is on the tenth day of the month of Rosh Hashanah, that the Day of Atonement was remembered. And it is on the tenth day of the month, that His witness shall end with the prophecy. "And I will grant authority to my two witnesses, and they will prophesy for twelve hundred and sixty days, clothed in sackcloth." (Revelation 11:3-4)

And the temple is held in the palms of His hands. For the tabernacle was covered with ten curtains of linen. (Exodus 26:1) And the tabernacle was held by ten pillars of gold. (Exodus 27:12) And the cherubim of the covenant was ten by ten cubits. (1 Kings 6:23-25) And in the inner courtyard were ten stands of bronze. (1 Kings 7:27) And upon the ten stands were ten basins of bronze. (1 Kings 7:38) And in the ten basins the priests washed the lamb. But when they entered the land, only one of ten could enter the city. (Nehemiah 11:1)

For the people have failed to heed the Lord's voice. "Surely all the men who have seen My glory and My signs which I performed in Egypt and in the wilderness, yet have put Me to the test these ten times and have not listened to My voice". (Number 14:22) And through the hands of the Son, ten would receive the bread of His Word, and ten would drink the blood of His sacrifice. Yet only one out of ten would return to give glory to God. ***And that one the Lord saved, for it was by FAITH that he was saved***. (Luke 17:11-19) For it is through faith that the Lamb gives us mercy. But on that day, only one of ten will be saved by the Lamb, for "in those days ten men from all the nations will grasp the garment of a Jew, saying, 'Let us go with you, for we have heard that God is with you." (Zechariah 8:23) For they shall not be judged by the Lamb, but by their lack of faith, He shall send them away. For as He promised "I will declare to them, 'I never knew you; DEPART FROM ME, YOU WHO PRACTICE LAWLESSNESS .'" (Matthew 7:23) And they shall return to the fire of judgment. And they shall not be judged by the Lamb, but they shall be judged by the Lion.

XI

The Number of Justice (NASB)

To Andrew the Prophet

Completed January 24, 2008

The 11th hour is upon us

"And about the **eleventh** hour he went out and found others standing around; and he said to them, 'Why have you been standing here idle all day long?' They said to him, 'Because no one hired us.' He said to them, 'You go into the vineyard too.' When evening came, the owner of the vineyard said to his foreman, 'Call the laborers and pay them their wages, beginning with the last group to the first.' When those hired about the **eleventh** hour came, each one received a denarius." Matthew 20:6-9

Only eleven of twelve remained

"For Jesus knew from the beginning who they were who did not believe, and who it was that would betray Him. And He was saying, 'For this reason I have said to you, that no one can come to Me unless it has been granted him from the Father.'" John 6:64-65

"'Why do you seek the living One among the dead? He is not here, but He has risen . Remember how He spoke to you while He was still in Galilee, saying that the Son of Man must be delivered into the hands of sinful men, and be crucified, and the third day rise again.' And they remembered His words, and returned from the tomb and reported all these things to the **eleven** and to all the rest." Luke 24:5-9

But the **eleven** disciples proceeded to Galilee, to the mountain which Jesus had designated. When they saw Him, they worshiped Him; but some were doubtful. And Jesus came up and spoke to them, saying, 'All authority has been given to Me in heaven and on earth. Go therefore and make disciples of all the nations, baptizing them in the name of the Father and the Son and the Holy Spirit, teaching them to observe all that I commanded you; and lo, I am with you always , even to the end of the age." Matthew 28:16-20

"And they got up that very hour and returned to Jerusalem, and found gathered together the **eleven** and those who were with them, saying, 'The Lord has really risen and has appeared to Simon.' They began to relate their experiences on the road and how He was recognized by them in the breaking of the bread." Luke 24:33-35

The eleventh hour is the hour of reckoning

"In the fortieth year, on the first day of the **eleventh** month, Moses spoke to the children of Israel, according to all that the Lord had commanded him to give to them." Deuteronomy 1:3

"In the **eleventh** year, in the month of Bul, which is the eighth month, the house was finished throughout all its parts and according to all its plans. So he was seven years in building it." 1 Kings 6:38

"On the twenty-fourth day of the **eleventh** month, which is the month Shebat, in the second year of Darius, the word of the Lord came to Zechariah the prophet, the son of Berechiah, the son of Iddo, as follows: I saw at night, and behold, a man was riding on a red horse, and he was standing among the myrtle trees which were in the ravine, with red, sorrel and white horses behind him. Then I said, 'My lord, what are these?' And the angel who was speaking with me said to me, 'I will show you what these are.' And the man who was standing among the myrtle trees answered and said, 'These are those whom the Lord has sent to patrol the earth.' So they answered the angel of the Lord who was standing among the myrtle trees and said, 'We have patrolled the earth, and behold, all the earth is peaceful and quiet.' Then the angel of the Lord said, 'O Lord of hosts, how long will You have no compassion for Jerusalem and the cities of Judah, with which You have been indignant these seventy years?' The Lord answered the angel who was speaking with me with gracious words, comforting words. So the angel who was speaking with me said to me, 'Proclaim, saying, 'Thus says the Lord of hosts, 'I am exceedingly jealous for Jerusalem and Zion. But I am very angry with the nations who are at ease; for while I was only a little angry, they furthered the disaster.' Therefore thus says the Lord, 'I will return to Jerusalem with compassion; My house will be built in it,' declares the Lord of hosts, 'and a measuring line will be stretched over Jerusalem.' Again, proclaim, saying, 'Thus says the Lord of hosts, 'My cities will again overflow with prosperity, and the Lord will again comfort Zion and again choose Jerusalem.''''' Zechariah 1:7-17

Les Propheties
(Century 1, Quatrain 63)

Michel de Nostredame

Pestilences extinguished, the world becomes smaller,
For a long time the lands will be inhabited in peace:
People will travel safely through the sky over land and wave:
Then wars will start anew.

The 11th Hour

September 11, 2001

The New York Trade Towers came down, and the books of the prophecy were opened. And we lie awake in the eleventh hour, for the fire and the judgment to come. But for those who are asleep in the field, that hour shall soon come to an end. And "about the eleventh hour he went out and found others standing around; and he said to them, 'Why have you been standing here idle all day long?'" (Matthew 20:6) For the eleventh hour is upon us, and the judgment of fire awaits. For in numerology eleven represents justice, and even the unrighteous judge knew that day would come. "Now will not God bring about justice for His elect who cry to Him day and night, and will He delay long over them? I tell you that He will bring about justice for them quickly. However, when the Son of Man comes, will He find faith on the earth?" (Luke 18:6-8) And how can men have such lack of faith, and ignore the symbolism of that fateful day?

On September 11, 2001 the North Tower was struck by Flight 11

"The bird of prey offers itself to the heavens"
Michel de Nostradame

Les Propheties

(Century 1, Quatrain 24)

The new city contemplating a condemnation
The bird of prey offers itself to the heavens.
After victory pardon to the captives,
Cremona and Mantua will have suffered great evils.

And what is Cremona? Luigi Cremona is an Italian mathematician who invented the Cremona diagram, an architectural method for calculating the force on a truss. For in the Twin Towers, the asbestos on the trusses had been removed, but the fireproofing of the trusses had not been replaced. And when the "birds of prey offered itself to the heavens", the heat caused the trusses to melt and collapse, and the lateral forces caused the Twin Towers to fall. For it was FEMA who concluded by the "Truss Theory", that it was truss failure that caused the Cremona to fall. (*World Trade Center Building Performance Study,* FEMA 2002, 2005*)*

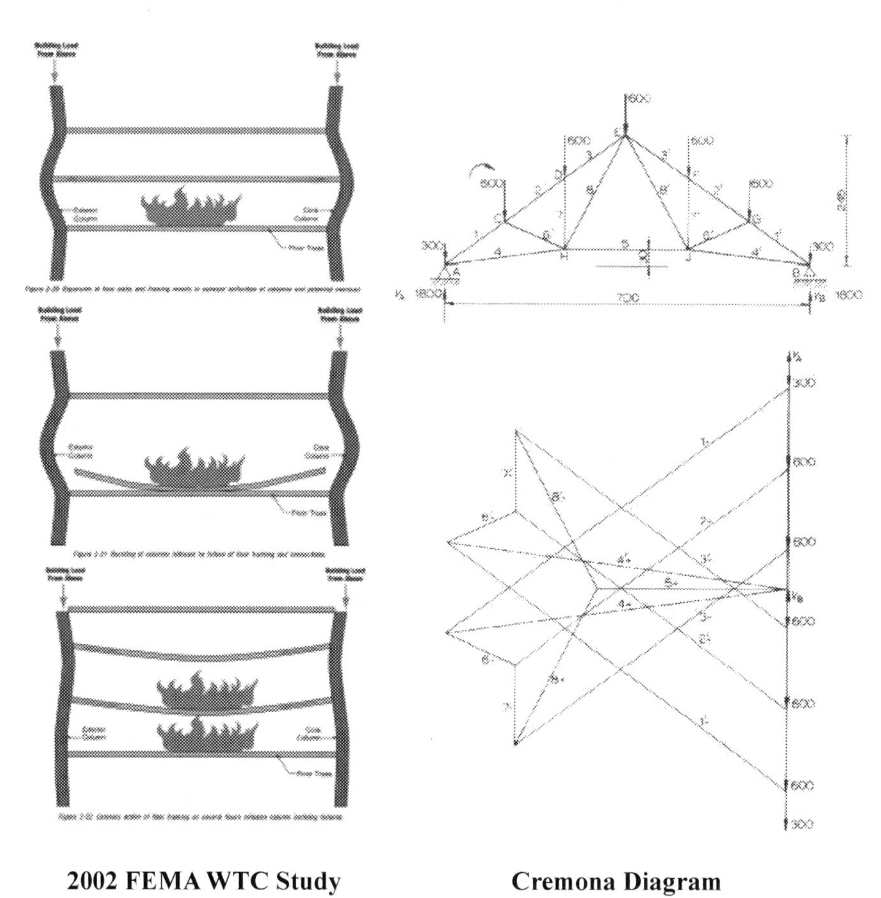

2002 FEMA WTC Study **Cremona Diagram**

And what is Mantua? Mantua is an ancient city in Italy, which looks like the vision of what Nostradamus saw on that Day.

And remember what the Lord said of the 11th hour. "When those hired about the eleventh hour came, each one received a denarius." (Matthew 20:9) Thus the time is still ripe for the workers, to labor for the Lord in the field. But for those found asleep in the field, they shall be sent to the fire.

For there were twelve who were appointed by the Lord. "And He went up on the mountain and summoned those whom He Himself wanted, and they came to Him. And He appointed twelve, so that they would be with Him and that He could send them out to preach." (Mark 3:13-14) But of the twelve who were appointed, one was left out. For eleven is derived from the German word *ainlif,* which means that "one is left out". For the one who was left out was a son of perdition. (John 6:64-65) "While I was with them, I was keeping them in Your name which You have given Me; and I guarded them and not one of them perished but the son of perdition, so that the Scripture would be fulfilled." (John 17:12) But after the Lord had arisen, the eleven returned to Jerusalem, to see Him breaking the bread. (Luke 24:33-35) And the Lord gave the eleven this commission. "Go therefore and make disciples of all the nations, baptizing them in the name of the Father and the Son and the Holy Spirit, teaching them to observe all that I commanded you; and lo, I am with you always, even to the end of the age." (Matthew 28:16-20)

For the eleventh hour is the final hour of reckoning. For in the eleventh month of the fortieth year, Moses instructed the people before his death in the desert. (Deuteronomy 1:3) And in the eleventh year of the temple's construction, the temple of God was complete. (1 Kings 6:38) And in the eleventh month of that year, Zecharias was told that the eleventh hour would come. "Thus says the Lord of hosts, 'I am exceedingly jealous for Jerusalem and Zion. But I am very angry with the nations who are at ease; for while I was only a little angry, they furthered the disaster.' Therefore thus says the Lord, 'I will return to Jerusalem with compassion; My house will be built in it,' declares the Lord of hosts, 'and a measuring line will be stretched over Jerusalem.' Again, proclaim, saying, 'Thus says the Lord of hosts, 'My cities will again overflow with prosperity, and the Lord will again comfort Zion and again choose Jerusalem.”””(Zecharias 1:7-17)

And in the year of two thousand and eleven, the final world war shall begin. And the bowls of wrath shall be poured out. And the fires of wrath shall burn flesh. For the eleventh hour is here, and the final hour shall end, and woe to those who are left in the field!

XII

The Twelve Tribes From All the Nations (NASB)

To Andrew the Prophet

Completed January 26, 2008

God created twelve tribes of Israel and Ishmael

“As for Ishmael, I have heard you; behold, I will bless him, and will make him fruitful and will multiply him exceedingly. He shall become the father of **twelve** princes, and I will make him a great nation.” Genesis 17:20

“Now there were **twelve** sons of Jacob—the sons of Leah: Reuben, Jacob’s firstborn, then Simeon and Levi and Judah and Issachar and Zebulun; the sons of Rachel: Joseph and Benjamin; and the sons of Bilhah, Rachel’s maid: Dan and Naphtali; and the sons of Zilpah, Leah’s maid: Gad and Asher. These are the sons of Jacob who were born to him in Paddan-aram. Genesis 35:22-26

“Moses wrote down all the words of the Lord. Then he arose early in the morning, and built an altar at the foot of the mountain with **twelve** pillars for the **twelve** tribes of Israel.” Exodus 24:4

The offerings were made in twelve

“Speak to the sons of Israel, and get from them a rod for each father’s household: **twelve** rods, from all their leaders according to their fathers’ households. You shall write each name on his rod.” Number 17:2

“The stones shall be according to the names of the sons of Israel: **twelve**, according to their names; they shall be like the engravings of a seal, each according to his name for the **twelve** tribes.” Exodus 28:21

“Then you shall take fine flour and bake **twelve** cakes with it; two-tenths of an ephah shall be in each cake.” Leviticus 24:5

"When they brought their offering before the Lord, six covered carts and **twelve** oxen, a cart for every two of the leaders and an ox for each one, then they presented them before the tabernacle...This was the dedication offering for the altar from the leaders of Israel when it was anointed: **twelve** silver dishes, **twelve** silver bowls, **twelve** gold pans, the **twelve** gold pans, full of incense, weighing ten shekels apiece, according to the shekel of the sanctuary, all the gold of the pans, 120 shekels; _all the oxen for the burnt offering **twelve** bulls, all the rams **twelve**, the male lambs one year old with their grain offering **twelve**, and the male goats for a sin offering **twelve**." Numbers 7:3, 84-87

The temple was built in twelve

"It stood on **twelve** oxen, three facing north, three facing west, three facing south, and three facing east; and the sea was set on top of them, and all their rear parts turned inward" 1 Kings 7:25

"**Twelve** lions were standing there on the six steps on the one side and on the other; nothing like it was made for any other kingdom." 1 Kings 10:20

"Now the altar hearth shall be **twelve** cubits long by **twelve** wide, square in its four sides." Ezekiel 43:16

God saves man in twelve

"Then they came to Elim where there were **twelve** springs of water and seventy date palms, and they camped there beside the waters." Exodus 15:27

"A woman who had had a hemorrhage for **twelve** years, and had endured much at the hands of many physicians, and had spent all that she had and was not helped at all, but rather had grown worse—after hearing about Jesus, she came up in the crowd behind Him and touched His cloak." Mark 5:25-27

"Immediately the girl got up and began to walk, for she was **twelve** years old. And immediately they were completely astounded." Mark 5:42

"Jesus answered, 'Are there not **twelve** hours in the day? If anyone walks in the day, he does not stumble, because he sees the light of this world.'" John 11:9

"when I broke the five loaves for the five thousand, how many baskets full of broken pieces you picked up?" They said to Him, '**Twelve**.'" Mark 8:19

The Word was handed to the twelve

"Jesus summoned His **twelve** disciples and gave them authority over unclean spirits, to cast them out, and to heal every kind of disease and every kind of sickness. Now the names of the **twelve** apostles are these: The first, Simon, who is called Peter, and Andrew his brother; and James the son of Zebedee, and John his brother; Philip and Bartholomew ; Thomas and Matthew the tax collector; James the son of Alphaeus, and Thaddaeus; Simon the Zealot , and Judas Iscariot, the one who betrayed Him." Matthew 10:1-4

"And Jesus said to them, "Truly I say to you, that you who have followed Me, in the regeneration when the Son of Man will sit on His glorious throne, you also shall sit upon **twelve** thrones, judging the **twelve** tribes of Israel." Matthew 19:28

And the New Kingdom will be twelve

"Now the altar hearth *shall be* **twelve** *cubits* long by **twelve** wide, square in its four sides." Ezekiel 43:16

"And I heard the number of those who were sealed, one hundred and forty-four thousand sealed from every tribe of the sons of Israel: From the tribe of Judah, **twelve** thousand were sealed, from the tribe of Reuben **twelve** thousand, from the tribe of Gad **twelve** thousand, from the tribe of Asher **twelve** thousand, from the tribe of Naphtali **twelve** thousand, from the tribe of Manasseh **twelve** thousand, from the tribe of Simeon **twelve** thousand, from the tribe of Levi **twelve** thousand, from the tribe of Issachar **twelve** thousand, from the tribe of Zebulun **twelve** thousand, from the tribe of Joseph **twelve** thousand, from the tribe of Benjamin, **twelve** thousand were sealed." Revelation 7:4-8

"It had a great and high wall, with **twelve** gates, and at the gates **twelve** angels; and names were written on them, which are the names of the **twelve** tribes of the sons of Israel." Revelation 21:12

"And the wall of the city had **twelve** foundation stones, and on them were the **twelve** names of the **twelve** apostles of the Lamb." Revelation 21:14

"On either side of the river was the tree of life, bearing **twelve** kinds of fruit, yielding its fruit every month; and the leaves of the tree were for the healing of the nations." Revelation 22:2

The Twelve Tribes and Kings

Twelve completes all things through time, for there are twelve months in a year. And there are twelve hours of light in a day, for as the Lord said "are there not twelve hours in the day? If anyone walks in the day, he does not stumble, because he sees the light of this world." (John 11:9) But soon the light of the day shall depart, for "behold, the day of the Lord is coming, cruel, with fury and burning anger, to make the land a desolation; and He will exterminate its sinners from it. For the stars of heaven and their constellations will not flash forth their light; the sun will be dark when it rises and the moon will not shed its light." (Isaiah 13:9-10)

And why does any nation claim favor from God? For God promised Ishmael the father of Islam, that "he shall become the father of twelve princes, and I will make him a great nation." (Genesis 17:20) And God promised Jacob the father of Judaism, that he would father the twelve tribes of Israel. (Genesis 35:22-26) And when Moses climbed up Mount Sinai, "an altar was built at the foot of the mountain with twelve pillars for the twelve tribes of Israel." (Exodus 24:4) Yet they continue to war to this day. But "do we not all have one father? Has not one God created us? Why do we deal treacherously each against his brother so as to profane the covenant of our fathers?" (Malachi 2:10)

And the offerings at the altar were in twelve. For from the twelfth rod of Aaron, the priests of the tabernacle were chosen. (Number 17:2) And the ephod of the priests held twelve precious stones. (Exodus

28:21) And the table of the offering held twelve loaves of bread. (Leviticus 24:5) And the offerings were brought by twelve leaders. And the offerings that they brought were twelve oxen, twelve silver dishes, twelve silver bowls, twelve gold pans with twelve shekels apiece, twelve bulls, twelve rams, twelve lambs, twelve goats, and twelve grain offerings. (Numbers 7:3, 84-87)

And the temple was built in the semblance of twelve tribes. For the sea of the temple was held up by twelve oxen. (1 Kings 7:25) And the steps to the throne were flanked by twelve lions. (1Kings 10:20) And the altar of the temple will be "twelve cubits long by twelve wide, square in its four sides." (Ezekiel 43:16) For the altar of the twelve is a testimony, to the sacrifices that His martyrs have made.

And the Lord shows us mercy in twelve. For when His people were exiled in the desert, He gave them twelve springs of water at Elim. (Exodus 15:27) And the Lord in His mercy sent His Son down to earth to save all mankind. For the Son healed the nations in twelve. For the woman who was hemorrhaging, in the twelfth year of her affliction, was healed with His robe. (Mark 5:25-27) And the daughter of the official was twelve years old, when she was arisen from the dead. (Mark 5:42) And when the masses were fed from the baskets, the Lord asked His disciples "when I broke the five loaves for the five thousand, how many baskets full of broken pieces did you pick up? They said to Him, '***<u>Twelve</u>***.'" (Mark 8:19)

For the Bread which is the Word of God, came down to earth to save all mankind. And He fed mankind with the Bread of His Word. And the Bread of the Word was broken, and the Bread was gathered in twelve baskets, and the baskets were handed to twelve apostles. And the twelve apostles were commissioned, to feed all of mankind with His Word. And He gathered His twelve apostles, and gave them authority over the powers of this world. (Matthew 10:1-4) Yet their authority over this world, is but a morsel of their authority that shall come, "Truly I say to you, that you who have followed Me, in the regeneration when the Son of Man will sit on His glorious throne, you also shall sit upon twelve thrones, judging the twelve tribes of Israel." (Matthew 19:28)

And their sacrifice for His Word was great indeed. For the altar of the temple was twelve by twelve cubits, in the semblance of the sacrifice of His bride. (Ezekiel 43:16) For "I heard the number of those who were sealed, one hundred and forty-four thousand sealed from every tribe of the sons of Israel." (Revelation 7:4-8) "Then one of the elders answered, saying to me, 'These who are clothed in the white robes, who are they, and where have they come from?' I said to him, 'My lord, you know.' And he said to me, 'These are the ones who come out of the great tribulation, and they have washed their robes and made them white in the blood of the Lamb. For this reason, they are before the throne of God; and they serve Him day and night in His temple; and He who sits on the throne will spread His tabernacle over them. They will hunger no longer, nor thirst anymore; nor will the sun beat down on them, nor any heat; for the Lamb in the center of the throne will be their shepherd, and will guide them to springs of the water of life; and God will wipe every tear from their eyes." (Revelation 7:13-17) For the twelve by twelve thousand are the Temple. For the true Temple of God is the Bride, who are the prophets and martyrs and saints. And on the walls of the city are twelve gates, and the gates are named for twelve tribes, and guarding the gates are twelve angels. (Revelation 21:12) And on the twelve gates are twelve stones, and on the twelve stones are the names of twelve apostles. For the twelve angels at the twelve gates are the twelve apostles of the Lord. (Revelation 21:14) And within the city walls will be a great river, and "on either side of the river was the tree of life, bearing twelve kinds of fruit, yielding its fruit every month; and the leaves of the tree were for the healing of the nations." (Revelation 22:2) For the twelve nations shall be healed and His kingdom shall be complete.

XIII

God is Water, the Sustenance of Life (NKJV)

To Andrew the Prophet

Completed August 22, 2007

God created water to sustain the earth

"And God called the dry land Earth, and the gathering together of the **waters** He called Seas. And God saw that it was good." Genesis 1:10

"Then God said, "Let the **waters** abound with an abundance of living creatures, and let birds fly above the earth across the face of the firmament of the heavens."" Genesis 1:20

"Now a river went out of Eden to **water** the garden, and from there it parted and became four riverheads." Genesis 2:10

"You visit the earth and water it, You greatly enrich it; The river of God is full of **water**; You provide their grain, For so You have prepared it." Psalms 65:9

God the Father is the Water of life, our Sustenance

"Jesus answered, "Most assuredly, I say to you, unless one is born of **water** and the Spirit, he cannot enter the kingdom of God."" John 3:5

"He who believes in Me, as the Scripture has said, out of his heart will flow rivers of living **water**." John 7:38

"And there are three that bear witness on earth: the Spirit, the **water**, and the blood; and these three agree as one."1 John 5:8

The Water will dwell amongst us forever

"Jesus answered and said to her, "Whoever drinks of this **water** will thirst again, "but whoever drinks of the water that I shall give him will never thirst. But the water that I shall give him will become in him a fountain of water springing up into everlasting life.""" John 4:13

"And He said to me, "It is done! I am the Alpha and the Omega, the Beginning and the End. I will give of the fountain of the **water** of life freely to him who thirsts."" Revelation 21:6

"And he showed me a pure river of **water** of life, clear as crystal, proceeding from the throne of God and of the Lamb. In the middle of its street, and on either side of the river, was the tree of life, which bore twelve fruits, each tree yielding its fruit every month. And the leaves of the tree were for the healing of the nations." Revelation 22:1

The Father is Water

Symbolism is allegorical and beautiful. And God gave us symbols to represent, that which we cannot understand, and that is the infinite spectrum of the Father, and the heavens and the earth He created. And the greatest symbolism is not in man's words, but the greatest symbolism is in creation itself. "For behold, He who forms mountains, And creates the wind, Who declares to man what his thought is, And makes the morning darkness, Who treads the high places of the earth--The Lord God of hosts is His name." (Amos 4:13)

And God created the water of the earth (Genesis 1:10), and from these waters all living things grew.(Genesis 1:20) And in the garden of Adam and Eve, the riverheads flowed with its sustaining life. (Genesis 2:10) For from that water the gardens grew grain, the bread that sustained their comfort and life. (Psalms 65:9)

And water can form complex hydrogen bonds, so biochemical processes can function fluently. Each of these processes can work independently, but without them functioning in coordination with each other, even the simplest life process is impossible. And the Father is sustenance and water, for He allows all things to work together in harmony. For "we know that all things work together for good to those who love God, to those who are the called according to His purpose." (Romans 8:28)

Water is a sustaining and cleansing fluid. It removes the debris and waste products from biological systems. And God is sustenance and cleansing. He is the sustenance of life, for all things exist through Him. And He is cleansing for mankind. For He sent His only Son to bring us the Word of God, and to shed His blood for man, that we may receive the Holy Spirit. For as the Lord said, we must be born of water, confessing our sins to Christ, and relying on the sustenance of the Father. And having cleansed our temples, we receive the Holy Spirit, that we may enter the kingdom of God. (John 3:5) And now that the Spirit dwells in us, "the rivers of living water" can flow from us. (John 7:38)

Water is supportive for any living organism. It allows the plants to support their stems. And our bodies consist of two-thirds water. Thus water gives our body structure, form, and strength. And the Father is our support and strength. He sustains our life and fulfills our needs. He gives us the spiritual guidance and strength, to weather the trials and storms. For the Father is the water which never fails. "The LORD will guide you always; he will satisfy your needs in a sun-scorched land and will strengthen your frame. You will be like a well-watered garden, like a spring whose waters never fail." (Isaiah 58:11)

Water exists in three different forms: vapor, water, and ice. Though they are the same substance, they exist in three different forms. Just as the Trinity is one God, but exists in three separate forms. And vapor is the breath of the Holy Spirit. "The Spirit of God has made me, And the breath of the Almighty gives me life." (Job 33:4) And ice is the rock of Christ, the cornerstone of the Church. "For they drank of that spiritual Rock that followed them, and that Rock was Christ" (1 Corinthians 10:4) And water is the sustenance from the Father, the sustenance and structure of creation. "You visit the earth and water it, You greatly enrich it; The river of God is full of water" (Psalms 65:9) "And there are three that bear witness on earth: the Spirit, the water (the Father), and the blood (the Son); and these three agree as one."(1 John 5:8)

And speaking of its structure, the molecule of water is a trinity. For it consists of three separate atoms: two hydrogens and an oxygen.

Oxygen symbolizes the Spirit, for the Spirit is the breath of life. And hydrogen symbolizes the Father, and hydrogen in the likeness of the Father is the Son. And the Son exists in unity with the Father, for as the Son said of the Father, "I and My Father are one." (John 10:30) And what if you fuse the Father and the Son? And what happens if you fuse hydrogen? Remember the holocaust of Hiroshima and Nagasaki? For what you get is the hydrogen bomb, the prototype of the modern day nuclear bomb. **The catastrophic release of energy is a result of man fusing the Father and the Son**!

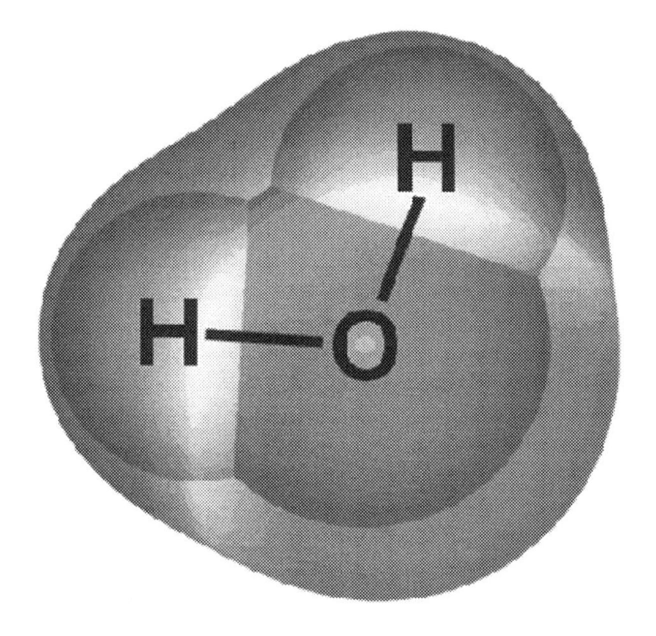

Figure 1

And now that we enter the end of this world, we will better understand the mysteries of the Trinity. For as we do know, it is impossible to divide one into three. (1 divided by 3 = 0.333333333.......∞). But for now we will have to rely, on what symbolism He has shown us.

But once the wrath of God, the payment of sins, and the reconciliation of man to the Father is complete, then all men will drink from the Water of life. And man will never be thirsty again, for the Water will flow freely for all. (Revelation 21:6) And man will never again suffer, for the Living Water shall heal the nations. (Revelation 22:1)

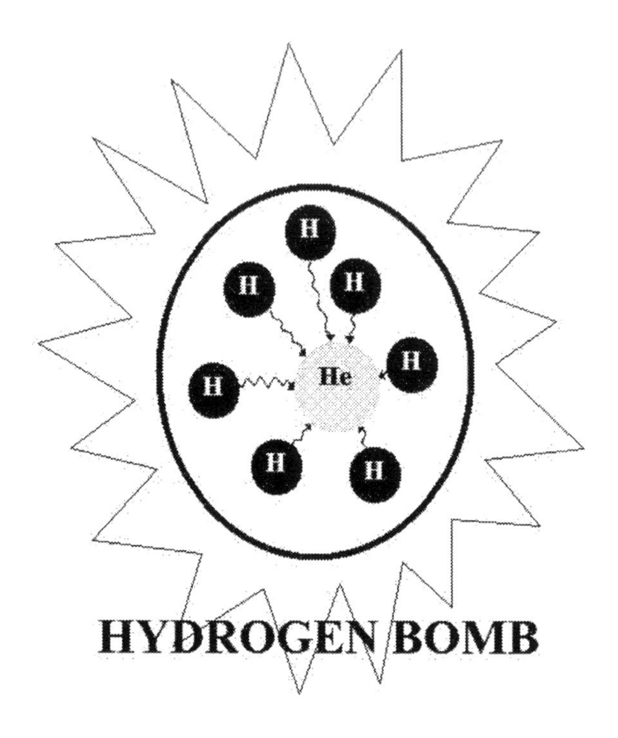

Figure 2

The First Hydrogen Bomb

The first hydrogen bomb was detonated on November 1, 1952 at the Marshall Islands. The blast produced a light greater than 1,000 suns and a heat wave that was felt over 50 kilometers away. Documentation from the Department of Energy states: "The immense ball of flame, cloud of dust, evaporated steel tower, melted sand for 1000 feet, 10 million tons of water rising out of the lagoon, waves subsiding from a height of 80 feet to seven feet in three miles, were all repeated in varying degrees, 43 times at Enewetak" Sound familiar? "And I saw something like a sea of glass mingled with fire, and those who have the victory over the beast, over his image and over his mark and over the number of his name, standing on the sea of glass, having harps of God." (Revelation 15:2)

Five islands in the atoll were completely vaporized by the testing. Nearly all the vegetation was destroyed, and half of the atoll remains uninhabitable. Thousands of Marshall Islanders were evacuated. An unknown number perished from starvation and radiation poisoning. But ultimately, the human toll is immeasurable.

XIV

Fourteen is the Day of the Lamb (NASB)

To Andrew the Prophet

Completed January 29, 2008

The kings came in divisions of fourteen

"So all the generations from Abraham to David are **fourteen** generations; from David to the deportation to Babylon, **fourteen** generations; and from the deportation to Babylon to the Messiah, **fourteen** generations." Matthew 1:17

The lamb was sacrificed in fourteen

"You shall keep it until the **fourteenth** day of the same month, then the whole assembly of the congregation of Israel is to kill it at twilight." Exodus 12:6

"In the first month, on the **fourteenth** day of the month at evening, you shall eat unleavened bread, until the twenty-first day of the month at evening." Exodus 12:18

"In the first month, on the **fourteenth** day of the month at twilight is the Lord's Passover. Then on the fifteenth day of the same month there is the Feast of Unleavened Bread to the Lord; for seven days you shall eat unleavened bread." Leviticus 23:5-7

The third temple of Ezekiel is in the image of the Kingdom to come

"In the twenty-fifth year of our exile, at the beginning of the year, on the tenth of the month, in the **fourteenth** year after the city was taken, on that same day the hand of the Lord was upon me and He brought me there." Ezekiel 40:1

"And these are the measurements of the altar by cubits (the cubit being a cubit and a handbreadth): the base shall be a cubit and the width a cubit, and its border on its edge round about one span; and this shall be the height of the base of the altar. From the base on the ground to the lower ledge shall be two cubits and the width one cubit; and from the smaller ledge to the larger ledge shall be four cubits and the width one cubit. The altar hearth shall be four cubits; and from the altar hearth shall extend upwards four horns. Now the altar hearth shall be twelve cubits long by twelve wide, square in its four sides. The ledge shall be **fourteen** cubits long by **fourteen** wide in its four sides, the border around it shall be half a cubit and its base shall be a cubit round about; and its steps shall face the east." Ezekiel 43:13-17

The Final Temple of God (Ezekiel 43)

The Day of the Sacrifice is the Fourteenth

The days of sacrifice are in fourteen. For from Abraham to David were fourteen generations. And from David until the deportation were fourteen generations. And from the deportation until the Messiah were fourteen generations. (Matthew 1:17)

And the sacrifice of His people was in fourteen. And on the fourteenth year of Jerusalem's fall, the Temple of God was revealed. (Ezekiel 43:13-17) And the altar measured twelve by twelve cubits, for the twelve tribes of twelve thousand martyrs. "And I heard the number of those who were sealed, one hundred and forty-four thousand sealed from every tribe of the sons of Israel." (Revelation 7:4) And the four horns are the four angels of God. "After this I saw four angels standing at the four corners of the earth, holding back the four winds of the earth, so that no wind would blow on the earth or on the sea or on any tree. And I saw another angel ascending from the rising of the sun, having the seal of the living God; and he cried out with a loud voice to the four angels to whom it was granted to harm the earth and the sea, saying, 'Do not harm the earth or the sea or the trees until we have sealed the bond-servants of our God on their foreheads.'" (Revelation 7:1-3) And the base of the temple is fourteen by fourteen cubits, for the multitude that comes from the tribulation. "After these things I looked, and behold, a great multitude which no one could count, from every nation and all tribes and peoples and tongues, standing before the throne and before the Lamb, clothed in white robes, and palm branches were in their hands; and they cry out with a loud voice, saying, 'Salvation to our God who sits on the throne, and to the Lamb.' And all the angels were standing around the throne and around the elders and the four living creatures; and they fell on their faces before the throne and worshiped God." (Revelation 7:9-11) But to those who forsake Him, they shall be the footstool of the Temple. For as "the Lord said to my Lord, 'Sit at my right hand, until I make Your enemies a footstool for Your feet.'" (Luke 20:42-43)

And the sacrifice of the Passover was in fourteen. For on the fourteenth day of the month, the unleavened bread was offered. (Exodus 12:18) And on the fourteenth day of the month, the blood of the lamb was offered, and the angel of death passed over, and the Lord would deliver His people. (Exodus 12:6) And on the fourteenth day of the month, they remember their deliverance through His hands. (Leviticus 23:5-7)

And the Lamb was sacrificed on the fourteenth. For on the fourteenth day of the month, on the eve of the Passover, He gathered His disciples at the Last Supper. "I have earnestly desired to eat this Passover with you before I suffer; for I say to you, I shall never again eat it until it is fulfilled in the kingdom of God." (Luke 22:15-16) And on the fourteenth day of the month, He offered the Bread of His Body. "This is My body which is given for you; do this in remembrance of Me." (Luke 22:19) And on the fourteenth day of the month, He offered the blood of the Lamb. "And in the same way He took the cup after they had eaten, saying, "This cup which is poured out for you is the new covenant in My blood." (Luke 22:20) And after the Passover of His Bread and His blood, the Lamb was sacrificed and died on the cross. And through the Fourteen Stations of the Cross, we remember His sacrifice for mankind. And His witness shall perish on the fourteenth day of the month. "When they have finished their testimony, the beast that comes up out of the abyss will make war with them, and overcome them and kill them." (Revelation 11:7)

The Fourteen Stations of the Cross
Via Dolorosa

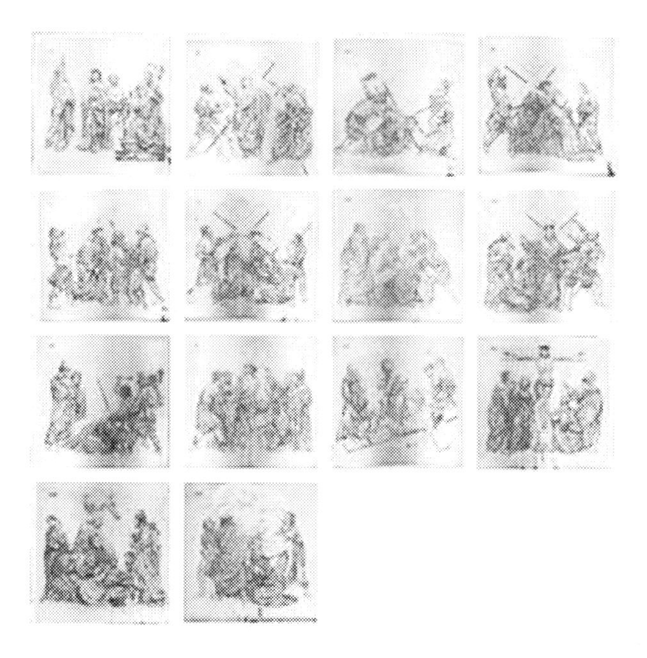

XV

The Kingdom was Handed to Mankind (NASB)

To Andrew the Prophet

Completed January 31, 2008

God delivers His people in fifteen

"Moses said to the people, 'Remember this day in which you went out from Egypt, from the house of slavery ; for by a powerful hand the Lord brought you out from this place. And nothing leavened shall be eaten. On this day in the month of Abib, you are about to go forth. It shall be when the Lord brings you to the land of the Canaanite, the Hittite, the Amorite, the Hivite and the Jebusite, which He swore to your fathers to give you, a land flowing with milk and honey, that you shall observe this rite in this month. For seven days you shall eat unleavened bread, and on the seventh day there shall be a feast to the Lord. Unleavened bread shall be eaten throughout the seven days; and nothing leavened shall be seen among you, nor shall any leaven be seen among you in all your borders. You shall tell your son on that day, saying, 'It is because of what the Lord did for me when I came out of Egypt.' And it shall serve as a sign to you on your hand, and as a reminder on your forehead, that the law of the Lord may be in your mouth; for with a powerful hand the Lord brought you out of Egypt. Therefore, you shall keep this ordinance at its appointed time from year to year." Exodus 13:3-10

"They journeyed from Rameses in the first month, on the **fifteenth** day of the first month; on the next day after the Passover the sons of Israel started out boldly in the sight of all the Egyptians,4 while the Egyptians were burying all their firstborn whom the Lord had struck down among them. The Lord had also executed judgments on their gods." Numbers 33:3-4

"Then they set out from Elim, and all the congregation of the sons of Israel came to the wilderness of Sin, which is between Elim and

Sinai, on the **fifteenth** day of the second month after their departure from the land of Egypt." Exodus 16:1

"But the Jews who were in Susa assembled on the thirteenth and the fourteenth of the same month, and they rested on the **fifteenth** day and made it a day of feasting and rejoicing." Esther 9:18

Israel's deliverance is celebrated on the fifteenth

"Then on the **fifteenth** day of the same month there is the Feast of Unleavened Bread to the Lord; for seven days you shall eat unleavened bread." Leviticus 23:6

"Speak to the sons of Israel, saying, 'On the **fifteenth** of this seventh month is the Feast of Booths for seven days to the Lord.' ...On exactly the **fifteenth** day of the seventh month, when you have gathered in the crops of the land, you shall celebrate the feast of the Lord for seven days, with a rest on the first day and a rest on the eighth day." Leviticus 23:34,39

"Then Mordecai recorded these events, and he sent letters to all the Jews who were in all the provinces of King Ahasuerus, both near and far, obliging them to celebrate the fourteenth day of the month Adar, and the **fifteenth** day of the same month, annually, because on those days the Jews rid themselves of their enemies, and it was a month which was turned for them from sorrow into gladness and from mourning into a holiday ; that they should make them days of feasting and rejoicing and sending portions of food to one another and gifts to the poor." Esther 9:20-22

The Hand of the Trinity and Man's Works

"God created man in His own image, in the image of God He created him; male and female He created them. God blessed them; and God said to them, 'Be fruitful and multiply, and fill the earth, and subdue it; and rule over the fish of the sea and over the birds of the sky and over every living thing that moves on the earth.'" (Genesis 1:27-28) Thus God handed His kingdom to mankind, that man may complete His good works. For "You make him to rule over the works of Your hands; You have put all things under his feet, all sheep and oxen, and also the beasts of the field, the birds of the heavens and the fish of the sea, whatever passes through the paths of the seas." (Psalms 8:6-8) But man in his ignorance and greed, has fornicated and destroyed the inheritance of the earth. Thus mankind and earth shall be destroyed. "For the Lord has a day of vengeance, a year of recompense for the cause of Zion. Its streams will be turned into pitch, and its loose earth into brimstone, and its land will become burning pitch. It will not be quenched night or day; its smoke will go up forever. From generation to generation it will be desolate; none will pass through it forever and ever." (Isaiah 34:8-10)

But it was on the fifteenth day that God would deliver His people. For it was on the fifteenth day of the month of Abib, that His people were delivered from Egypt. (Exodus 13:3-10/Numbers 33:3-4) And it was on the fifteenth day after the first month of their exile, that God delivered them manna from heaven. (Exodus 16:1) And it was on the fifteenth day of the month of Adar, that they celebrated their deliverance from Persia. (Esther 9:18)

And it is on the fifteenth day that His people remember their deliverance. For it is on the fifteenth day of the month of Nissan, that they remember their deliverance from Egypt. (*Passover and Feast of Unleavened Bread*) (Exodus 13:3-10/Leviticus 23:6) And it is on the fifteenth day of the month of Tishri, that they remember their deliverance from the desert. (*Sukkot and Feast of Booths)* (Leviticus 23:34,39) And it is on the fifteenth day of the month of Adar, that they remember their deliverance from Persia. *(Shushan Purim)* (Esther 9:20-22)

And the Son delivered mankind in fifteen. For He came down to earth to teach us the Word, and was crucified and died on the cross. And on the fifteenth day of the month, He descended into hell, that its prisoners may be set free. For by the hands of His sacrifice, the Father handed all things to the Son. For "the Father loves the Son and has given all things into His hand." (John 3:35) And now the Son sits at the right hand of the Father, "He was received up into heaven and sat down at the right hand of God." (Mark 16:19) And the Son protects His sheep from the evil one, for He holds their lives in His hand. "My sheep hear My voice, and I know them, and they follow

Me; and I give eternal life to them, and they will never perish; and no one will snatch them out of My hand." (John 10:27-28) And His sheep have been appointed by the Shepherd, for the Spirit works all things through their hands. For "what is man, that You remember him? Or the son of man, that You are concerned about him? You have made him for a little while lower than the angels; You have crowned him with glory and honor, and have appointed him over the works of Your hands" (Hebrews 2:6-7)

And when the works of the Spirit are complete, all things shall be handed back to the Father. And when the last trumpet is blown, the mystery of God shall be complete. "Then the angel whom I saw standing on the sea and on the land lifted up his right hand to heaven, and swore by Him who lives forever and ever, who created heaven and the things in it, and the earth and the things in it, and the sea and the things in it, that there will be delay no longer, but in the days of the voice of the seventh angel, when he is about to sound, then the mystery of God is finished, as He preached to His servants the prophets." (Revelation 10:5-7)

The Three Hands of God

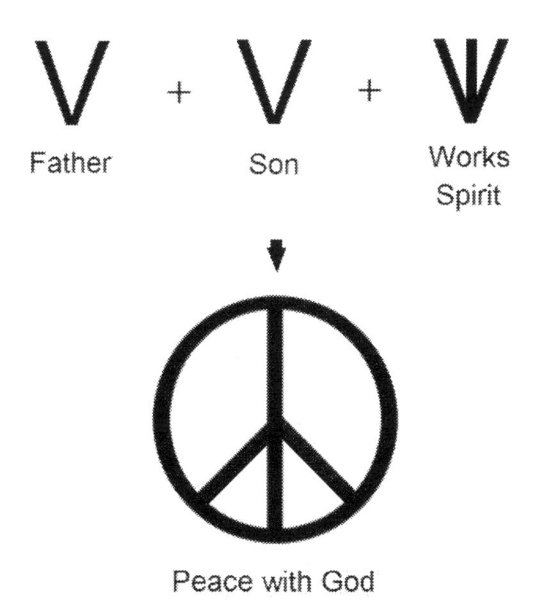

"For there will be peace for the seed: the vine will yield its fruit, the land will yield its produce and the heavens will give their dew; and I will cause the remnant of this people to inherit all these things. It will come about that just as you were a curse among the nations, O house of Judah and house of Israel, so I will save you that you may become a blessing. Do not fear; let your hands be strong." (Zechariah 8:12-13)

XVI

Hydrogen: The Father and the Son are One (NASB)

To Andrew the Prophet

Completed January 26, 2008

Hydrogen has the atomic number of one, and constitutes over ninety percent of this universe. And God is One for He is the Creator of the universe. "Lift up your eyes on high And see who has created these stars, the One who leads forth their host by number, He calls them all by name; because of the greatness of His might and the strength of His power, not one of them is missing." (Isaiah 40:26) And hydrogen exists as a diatomic molecule. And the Father and the Son are One, for as the Son said "I and the Father are one." (John 10:30)

Although hydrogen is the most abundant element in the universe, it is only the third most common element found on earth. For the Father dwells above the earth in the heavens, for as we do pray, "Our Father who art in heaven". (Matthew 6:9) For this world is not ruled by the Son, but is under the dominion of its ruler who is Satan. For as the Lord said, "I will not speak much more with you, for the ruler of the world is coming, and he has nothing in Me." (John 14:30)

And hydrogen can be highly explosive and flammable. And the day will soon come, when God's anger is rekindled, for He "will make the heavens tremble, and the earth will be shaken from its place at the fury of the Lord of hosts in the day of His burning anger." (Isaiah 13:13)

And man in his ignorance has stirred up the Lord's wrath. For the pot believes that he is the Potter. But "woe to the one who quarrels with his Maker - an earthenware vessel among the vessels of earth! Will the clay say to the potter , 'What are you doing?' Or the thing you are making say, 'He has no hands'?" (Isaiah 45:9) For man has altered the diatomic structure of hydrogen, to form the isotope of hydrogen which is tritium (^3H). And what is tritium? Tritium is a radioactive hydrogen isotope, which contains a proton and two neutrons. And tritium is utilized in the production of nuclear weapons. For tritium is an essential component of a fusion reaction, which boosts the efficiency of a fission bomb. And in January of 2008, Iran produced its first supply of tritium. And the rest shall be history.

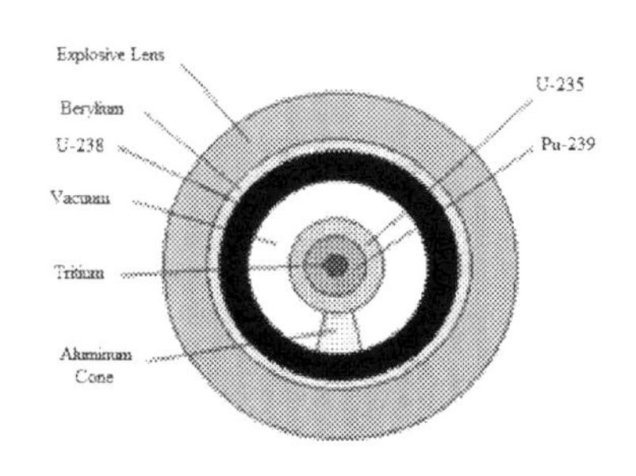

Fission Bomb

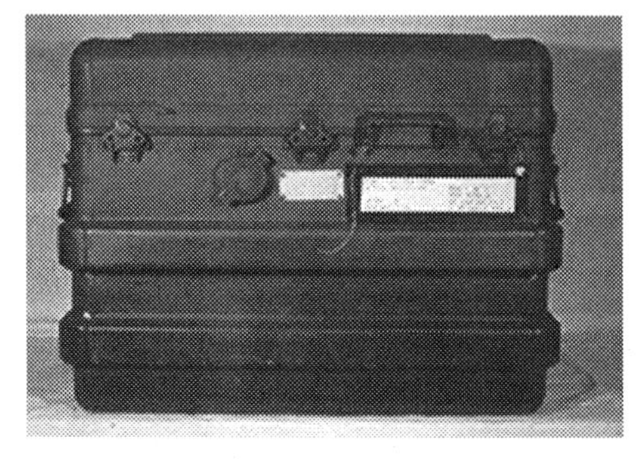

RA115: Prototype of USSR Suitcase Nuclear Arsenal

XVII

Oxygen is the Breath of the Spirit (NASB)

To Andrew the Prophet

Completed January 25, 2008

Oxygen is a colorless chemical represented by the chemical symbol **O.** It is derived from the Greek root *oxys* for "sharp" and *genesis* for "beginning". For the breath of the Spirit is a sharp sword, for "the sword of the Spirit, is the word of God." (Ephesians 6:17) And the breath of the Spirit dwelt in man from the beginning, for "the Holy Spirit fell upon them just as He did upon us at the beginning". (Acts 11:15) And oxygen abounds throughout all of the earth. And the Spirit of God dwells throughout the whole earth, for it is the Spirit of the earth, for "your voice will also be like that of a spirit from the ground, and your speech will whisper from the dust." (Isaiah 29:4) And oxygen enriches and strengthens our inner body. For the Spirit works through our inner body, "according to the riches of His glory, to be strengthened with power through His Spirit in the inner man." (Ephesians 3:16)

Oxygen is reactive and is integrated into most compounds. And as the time draws near, the Spirit shall pour out on all mankind, for "it will come about after this that I will pour out My Spirit on all mankind". (Joel 2:28) But oxygen can also be toxic to anaerobic (oxygen-less) organisms. And the Spirit shall bring forth His justice, on those without His Spirit, for "I have put My Spirit upon Him; He will bring forth justice to the nations." (Isaiah 42:1)

But at the genesis of earth's creation, oxygen was absent from the atmosphere. But soon after anaerobic bacteria began to produce oxygen, and thus oxygen was formed on the surface of the seas. And "in the beginning God created the heavens and the earth. The earth was formless and void, and darkness was over the surface of the deep, and the Spirit of God was moving over the surface of the waters." (Genesis 1:1-2) But the production of oxygen in the atmosphere, would eventually destroy most of earth's anaerobic organisms. And with the great flood, "all flesh that moved on the earth perished, birds and cattle and beasts and every swarming thing that swarms upon the earth, and all mankind; of all that was on the dry land, all in whose nostrils was the breath of the spirit of life, died." (Genesis 7:21-22)

And through photosynthesis and cellular respiration, the evolution of plants and animals would flourish. Photosynthesis is the process by which carbon dioxide, water, and sunlight is consumed by plants, algae, and cyanobacteria and is converted into oxygen and glucose.

$$6\,CO_2 + 6\,H_2O + \text{light} \rightarrow C_6H_{12}O_6 + 6\,O_2$$

And as we know, the Father is Water, the Son is Light, the Spirit is Breath, and glucose is the fruit of the Spirit (from *Signs, Science, and Symbols of the Prophecy*). Thus:

Spirit + Father + Son ➔ Fruit + Spirit

And His children produce the fruits of the Spirit. And "You will know them by their fruits. Grapes are not gathered from thorn bushes nor figs from thistles, are they? So every good tree bears good fruit, but the bad tree bears bad fruit. A good tree cannot produce bad fruit, nor can a bad tree produce good fruit." (Matthew 7:16-18) For "the fruit of the Spirit is love, joy, peace, patience, kindness, goodness, faithfulness, gentleness, self-control". (Galatians 5:22-23)

And there is a duality to photosynthesis, for that dual process is cellular respiration. And cellular respiration is the process by which aerobic organisms utilize oxygen and glucose to generate ATP (adenosine triphosphate). And ATP is a molecule which supplies energy to organisms at the cellular level.

$$C_6H_{12}O_6 + 6O_2 \rightarrow 6\,CO_2 + 6H_2O + ATP$$

And as we know, glucose is the fruit of the spirit, oxygen and CO2 are the breath of the spirit, and water is the sustenance of the Father. And ATP is the life of mankind. Thus:

Fruit + Spirit ➔ Spirit + Father + life of man

Thus, the fruits of the Spirit are not for God's benefit, but for the benefit of mankind. For "it is the Spirit who gives life; the flesh profits nothing; the words that I have spoken to you are spirit and are life. But there are some of you who do not believe." (John 6:63-64)

And now is the time to believe. For as the Word has promised, the wrath and the holocaust shall occur, and every living organism in the ocean shall die. "The second angel poured out his bowl into the sea, and it became blood like that of a dead man; and every living thing in the sea died." (Revelation 16:3) For the algae and the bacteria of the sea, coexist in a very delicate and sensitive ecosystem. But when the nuclear holocaust arrives, the ocean temperatures shall rise. And when the ocean temperatures rise, *K. brevis* which is a red algae in the ocean will bloom, and will create a deadly growth known as a "red tide". And these "red tides" produce neurotoxins which kill every living creature in its path. (bacteria, fish, algae, crustaceans, mammals, etc.)

Red Tide due to the organism *Karenia Brevis*

And as we know, the bacteria and algae of the sea produces more than two-thirds of the oxygen of the earth. Thus as the living organisms of the oceans and seas die, then the earth's oxygen shall be removed. And at the end the Spirit of God shall be removed, "then the dust will return to the earth as it was, and the spirit will return to God who gave it." (Ecclesiastes 12:7)

For the Lord had warned us, not to forsake the Spirit. "Therefore I say to you, any sin and blasphemy shall be forgiven people, but blasphemy against the Spirit shall not be forgiven. Whoever speaks a word against the Son of Man, it shall be forgiven him; but whoever speaks against the Holy Spirit, it shall not be forgiven him, either in this age or in the age to come." (Matthew 12:31-32)

For oxygen reacts with the metals of the earth. And the Spirit judges the works of all men. For as the Lord had forewarned us, "do not store up for yourselves treasures on earth, where moth and rust destroy, and where thieves break in and steal. But store up for yourselves treasures in heaven, where neither moth nor rust destroys, and where thieves do not break in or steal; for where your treasure is, there your heart will be also." (Matthew 6:19-21) For when oxygen reacts with metal, the metal is oxidized to rust, and the rust is consumed in fire. And He shall take the treasures of man and "set it empty on its coals so that it may be hot and its bronze may glow and its filthiness may be melted in it, its rust consumed. She has wearied Me with toil, yet her great rust has not gone from her; let her rust be in the fire!" (Ezekiel 24:11-12)

But the greatest woe is to those who held treasures in oil. For oil is a combustible and flammable compound. And as foretold "he who loves pleasure will become a poor man; he who loves wine and oil will not become rich." (Proverbs 21:17)

The equation for the combustion of oil (hydrocarbons) is:

$$CH4 + 2\,O_2 \rightarrow CO_2 + 2\,H_20 + \text{Heat}$$

For hydrocarbons were created without oxygen. And the sons of perdition were created without Spirit. "And all the arrogant and every evildoer will be chaff." (Malachi 4:1)

Thus:

Chaff + Spirit ➔ Breath of the Spirit + Father + Fire

"'Now I will arise,' says the Lord, 'Now I will be exalted, now I will be lifted up. You have conceived chaff, you will give birth to stubble; My breath will consume you like a fire.'" (Isaiah 33:10-11)

XVIII

God Adds Up All Things (NASB)

To Andrew the Prophet

Completed February 1, 2008

God allowed His people restitution for their sins

"He shall make restitution for that which he has sinned against the holy thing, and shall **add** to it a fifth part of it and give it to the priest. The priest shall then make atonement for him with the ram of the guilt offering, and it will be forgiven him." Leviticus 5:16

"But if a man eats a holy gift unintentionally, then he shall **add** to it a fifth of it and shall give the holy gift to the priest." Leviticus 22:14

"When a man or woman commits any of the sins of mankind, acting unfaithfully against the Lord, and that person is guilty, then he shall confess his sins which he has committed, and he shall make restitution in full for his wrong and **add** to it one-fifth of it, and give it to him whom he has wronged." Numbers 5:6-7

"Then on the sabbath day two male lambs one year old without defect, and two-tenths of an ephah of fine flour mixed with oil as a grain offering, and its drink offering: This is the burnt offering of every sabbath in **addition** to the continual burnt offering and its drink offering." Numbers 28:9-10

God allowed man to redeem their offering

"If, however, it is any unclean animal of the kind which men do not present as an offering to the Lord, then he shall place the animal before the priest. The priest shall value it as either good or bad; as you, the priest, value it, so it shall be. But if he should ever wish to redeem it, then he shall **add** one-fifth of it to your valuation." Leviticus 27:11-13

"Now if a man consecrates his house as holy to the Lord, then the priest shall value it as either good or bad; as the priest values it, so it shall stand. Yet if the one who consecrates it should wish to redeem his house, then he shall **add** one-fifth of your valuation price to it, so that it may be his." Leviticus 27:14-15

"If the one who consecrates it should ever wish to redeem the field, then he shall **add** one-fifth of your valuation price to it, so that it may pass to him." Leviticus 27:19

"Thus all the tithe of the land, of the seed of the land or of the fruit of the tree, is the Lord's; it is holy to the Lord. If, therefore, a man wishes to redeem part of his tithe, he shall **add** to it one-fifth of it." Leviticus 27:30-31

And God adds to the promises He has given

"If the Lord your God enlarges your territory, just as He has sworn to your fathers, and gives you all the land which He promised to give your fathers— if you carefully observe all this commandment which I command you today, to love the Lord your God, and to walk in His ways always—then you shall **add** three more cities for yourself, besides these three. So innocent blood will not be shed in the midst of your land which the Lord your God gives you as an inheritance, and bloodguiltiness be on you." Deuteronomy 19:8-10

"Return and say to Hezekiah the leader of My people, 'Thus says the Lord, the God of your father David, 'I have heard your prayer, I have seen your tears; behold, I will heal you. On the third day you shall go up to the house of the Lord. I will **add** fifteen years to your life, and I will deliver you and this city from the hand of the king of Assyria; and I will defend this city for My own sake and for My servant David's sake.''" 2 Kings 20:5-6

But instead man added to God's wrath

"You shall not marry a woman in **addition** to her sister as a rival while she is alive, to uncover her nakedness." Leviticus 18:18

"Woe to those who **add** house to house and join field to field, Until there is no more room, so that you have to live alone in the midst of the land!" Isaiah 5:8

"Enough of all your abominations, O house of Israel, when you brought in foreigners, uncircumcised in heart and uncircumcised in flesh, to be in My sanctuary to profane it, even My house, when you offered My food, the fat and the blood; for they made My covenant void—this in **addition** to all your abominations." Ezekiel 44:6-7

"So the Lord's anger burned against Israel, and He made them wander in the wilderness forty years, until the entire generation of those who had done evil in the sight of the Lord was destroyed. Now behold, you have risen up in your fathers' place, a brood of sinful men, to **add** still more to the burning anger of the Lord against Israel. For if you turn away from following Him, He will once more abandon them in the wilderness, and you will destroy all these people." Numbers 32:13-15

Nothing can be added to God's Word

"You shall not **add** to the word which I am commanding you, nor take away from it, that you may keep the commandments of the Lord your God which I command you. Your eyes have seen what the Lord has done in the case of Baal-peor, for all the men who followed Baal-peor, the Lord your God has destroyed them from among you. But you who held fast to the Lord your God are alive today, every one of you." Deuteronomy 4:2-4

"Every word of God is tested; He is a shield to those who take refuge in Him. Do not **add** to His words Or He will reprove you, and you will be proved a liar." Proverbs 30:5-6

And nothing can be added to God's plans

"I know that everything God does will remain forever; there is nothing to **add** to it and there is nothing to take from it, for God has so worked that men should fear Him. That which is has been already and that which will be has already been, for God seeks what has passed by." Ecclesiastes 3:14-15

"Look at the birds of the air, that they do not sow, nor reap nor gather into barns, and yet your heavenly Father feeds them. Are you not worth much more than they? And who of you by being worried can **add** a single hour to his life ?" Matthew 6:26-27

And now mankind must pay for the sum of their sins

"'Woe to the rebellious children,' declares the Lord, 'Who execute a plan, but not Mine, And make an alliance, but not of My Spirit, In order to **add** sin to sin'" Isaiah 30:1

"Thus says the Lord of hosts, the God of Israel, '**Add** your burnt offerings to your sacrifices and eat flesh. For I did not speak to your fathers, or command them in the day that I brought them out of the land of Egypt, concerning burnt offerings and sacrifices. But this is what I commanded them, saying, 'Obey My voice, and I will be your God, and you will be My people; and you will walk in all the way which I command you, that it may be well with you.' Yet they did not obey or incline their ear, but walked in their own counsels and in the stubbornness of their evil heart, and went backward and not forward." Jeremiah 7:21-24

The Summation of Man's Sins

Addition is derived from the Latin word *addere,* which means "to supply" or "to give". For the Father is willing to give to mankind, and man is willing to take from the Father. And as a father gives gifts to his children, so does the Father give gifts to His children. For as the Lord said, "if you then, being evil, know how to give good gifts to your children, how much more will your heavenly Father give the ***Holy Spirit*** to those who ask Him?" (Luke 11:13) But rather than asking for that which is good, man has asked for that which is evil. Thus "you ask and do not receive, because you ask with wrong motives, so that you may spend it on your pleasures." (James 4:3)

And due to man's nature all men tend to sin, yet God gave His people restitution for their sins, by allowing them to add to their sacrifices. And when they would sin against a holy thing, they could add one fifth to their offerings. (Leviticus 5:16) And when a man wrongfully took what was not his, he could add one fifth to his sacrifice. (Leviticus 22:14) And if any person would sin against another, he could one fifth to his offering. (Numbers 5:6-7) And on the Sabbath day an offering was added, to the burnt and drink offerings at the table. (Numbers 28:9-10)

And God ultimately requires nothing of man, and allows man to redeem what's been offered. For the unclean animals could be redeemed, by adding one-fifth to their estimated value. (Leviticus 27:11-13) And their consecrated homes could be redeemed, by adding one-fifth to its calculated value. (Leviticus 27:14-15) And the field that was consecrated could be redeemed, by adding one-fifth to the value of its price. (Leviticus 27:19) And a tithe of the land could be redeemed, by adding one-fifth to the value of its harvest. (Leviticus 27:30-31)

And God generously adds to the promises He has given. For He promised to increase the inheritance of His people, if they carefully observed His command, not to shed innocent blood on the land. (Deuteronomy 19:8-10) And to Hezekiah the king of His people, He promised to add fifteen years to his life, if he prayed and petitioned to the Father. (2 Kings 20:5-6) But His people refused to obey His commands, and shed innocent blood on the soil of the land. Thus the Lord would add vengeance and anger, to the sins they committed through their hands.

For they forsook His commands and had many wives. (Leviticus 18:18) And in greed and idolatry, they hoarded their land. (Isaiah 5:8) And they allowed foreign nations to desecrate the temple. (Ezekiel 44:6-7) And the Lord would abandon His people for their sins. "Now behold, you have risen up in your fathers' place, a brood of sinful men, to ***add*** still more to the burning anger of the Lord against Israel. For if you turn away from following Him, He will once more abandon them in the wilderness, and you will destroy all these people." (Numbers 32:13-15)

And the Word is not to be added to, for the Word of the Lord is complete. For "every word of God is tested; He is a shield to those who take refuge in Him. Do not ***add*** to His words or He will reprove you, and you will be proved a liar." (Proverbs 30:5-6) And He promises that by heeding His Words, that all of the people would have life. (Deuteronomy 4:2-4)

And addition is identified by zero, for when zero is added to any number, the quantity of that number remains the same. And man shall add not one thing, but that which is wrought by God's hands. For as Solomon said, "I know that everything God does will remain forever; there is nothing to ***add*** to it and there is nothing to take from it, for God has so worked that men should fear Him. That which is has been already and that which will be has already been, for God seeks what has passed by." (Ecclesiastes 3:14-15) But man in his ignorance knows no God. "Look at the birds of the air, that they do not sow, nor reap nor gather into barns, and yet your heavenly Father feeds them. Are you not worth much more than they? And who of you by being worried can ***add*** a single hour to his life?" (Matthew 6:26-27)

But the time to be worried is now, for the final hour of judgment is near. And all of man's works shall be summed up. For the Latin word for sum is *summa* for "the highest". And the Highest One will sum all the sins man has added. For "the ***sum*** of Your word is truth, and every one of Your righteous ordinances is everlasting." (Psalms 119:16) But "'woe to the rebellious children,' declares the Lord, 'Who execute a plan, but not Mine, and make an alliance, but not of My Spirit, in order to ***add*** sin to sin'" (Isaiah 30:1) For the Lord did not ask for burnt offerings, but for man to obey His commands. But man has refused to obey and to fear Him. "Thus says the Lord of hosts, the God of Israel, '***Add*** your burnt offerings to your sacrifices and eat flesh.'" (Jeremiah 7:21) And man shall eat their burnt offerings in the fire.

*Cogito, ergo **sum***

I think, therefore I am

Rene Descartes

Dominus cogito, ergo sum
God thinks, therefore I am

$\alpha\tau\pi$

XIX

God Shall Remove All Things that Oppose Him (NASB)

To Andrew the Prophet

Completed February 3, 2008

God commands us to remove anything between us and Him

"Then He said, 'Do not come near here; **remove** your sandals from your feet, for the place on which you are standing is holy ground.'" Exodus 3:5

"The captain of the Lord's host said to Joshua, '**Remove** your sandals from your feet, for the place where you are standing is holy.' And Joshua did so.'" Joshua 5:15

"Then Samuel spoke to all the house of Israel, saying, 'If you return to the Lord with all your heart, **remove** the foreign gods and the Ashtaroth from among you and direct your hearts to the Lord and serve Him alone; and He will deliver you from the hand of the Philistines.'" 1 Samuel 7:3

"Wash yourselves, make yourselves clean; **remove** the evil of your deeds from My sight. Cease to do evil" Isaiah 1:16

"**Remove** the false way from me, and graciously grant me Your law." Psalms 119:29

God ordered the priests to remove the imperfections from their offerings

"And the two kidneys with the fat that is on them, which is on the loins, and the lobe of the liver, which he shall **remove** with the kidneys." Leviticus 3:4

"Now a man who is clean shall gather up the ashes of the heifer and deposit them outside the camp in a clean place, and the congregation of the sons of Israel shall keep it as water to **remove** impurity; it is purification from sin." Numbers 19:9

"Then he shall **remove** all its fat, just as the fat was **remove**d from the sacrifice of peace offerings; and the priest shall offer it up in smoke on the altar for a soothing aroma to the Lord. Thus the priest shall make atonement for him, and he will be forgiven." Leviticus 4:31

"A quart of wheat for a denarius , and three quarts of barley for a denarius ; and do not damage the **oil** and the **wine**. " Revelation 6:6

God removes punishment when He pleases

"Then Pharaoh called for Moses and Aaron and said, 'Entreat the Lord that He **remove** the frogs from me and from my people; and I will let the people go, that they may sacrifice to the Lord.'" Exodus 8:8

"Now therefore, please forgive my sin only this once, and make supplication to the Lord your God, that He would only **remove** this death from me." Exodus 10:17

"But you shall serve the Lord your God, and He will bless your bread and your water; and I will **remove** sickness from your midst." Exodus 23:25

"The Lord sent fiery serpents among the people and they bit the people, so that many people of Israel died. So the people came to Moses and said, 'We have sinned, because we have spoken against the Lord and you; intercede with the Lord, that He may **remove** the serpents from us." And Moses interceded for the people. Then the Lord said to Moses, 'Make a fiery serpent, and set it on a standard; and it shall come about, that everyone who is bitten, when he looks at it, he will live.' And Moses made a bronze serpent and set it on the standard; and it came about, that if a serpent bit any man, when he looked to the bronze serpent, he lived.'" Numbers 21:6-9

"**Remove** Your plague from me; because of the opposition of Your hand I am perishing." Psalms 39:10

But man has removed their hearts from the Lord

“Circumcise yourselves to the Lord and **remove** the foreskins of your heart, Men of Judah and inhabitants of Jerusalem, Or else My wrath will go forth like fire And burn with none to quench it, Because of the evil of your deeds.” Jeremiah 4:4

“Then the Lord said, ‘Because this people draw near with their words and honor Me with their lip service, but they **remove** their hearts far from Me, And their reverence for Me consists of tradition learned by rote’” Isaiah 29:13

“So now let Me tell you what I am going to do to My vineyard: I will **remove** its hedge and it will be consumed; I will break down its wall and it will become trampled ground.” Isaiah 5:5

“I will also turn My hand against you, and will smelt away your dross as with lye and will **remove** all your alloy.” Isaiah 1:25

“Thus says the Lord God, ‘**Remove** the turban and take off the crown; this will no longer be the same. Exalt that which is low and abase that which is high.’” Ezekiel 21:26

“Then all the princes of the sea will go down from their thrones, **remove** their robes and strip off their embroidered garments. They will clothe themselves with trembling; they will sit on the ground, tremble every moment and be appalled at you.” Ezekiel 26:16

“Therefore thus says the Lord, ‘Behold, I am about to **remove** you from the face of the earth. This year you are going to die, because you have counseled rebellion against the Lord.’” Jeremiah 28:16

His servants have returned to claim back His kingdom

“‘For behold, the stone that I have set before Joshua; on one stone are seven eyes. Behold, I will engrave an inscription on it,’ declares the Lord of hosts, ‘and I will **remove** the iniquity of that land in one day.’” Zechariah 3:9

“When they come there, they will **remove** all its detestable things and all its abominations from it.” Ezekiel 11:18

"Go through, go through the gates, clear the way for the people; build up, build up the highway, **remove** the stones, lift up a standard over the peoples." Isaiah 62:10

In the end, God will remove the last enemy which is Death

"Moreover, I will give you a new heart and put a new spirit within you; and I will **remove** the heart of stone from your flesh and give you a heart of flesh." Ezekiel 36:26

"He will swallow up death for all time, And the Lord God will wipe tears away from all faces, And He will **remove** the reproach of His people from all the earth; For the Lord has spoken." Isaiah 25:8

"Then you will call, and the Lord will answer; you will cry, and He will say, 'Here I am.' If you **remove** the yoke from your midst, the pointing of the finger and speaking wickedness" Isaiah 58:9

All God's Enemies Shall be Removed

In the beginning all things were added by God, and at the end all things shall be subtracted by God. For as the Lord says, "I will completely remove all things from the face of the earth." (Zephaniah 1:2) For the function of subtraction is the inverse of addition, and the solution of the equation is the difference. For Adam and Eve chose to follow Satan's rule, thus all men became "his slaves so that they may learn the ***difference*** between My service and the service of the kingdoms of the countries." (2 Chronicles 12:8).

And the derivation of subtraction is the Latin word *subtractus,* which means "to remove from beneath". And God commands us to remove any evil, to remove all that comes from beneath. For when Moses walked on holy ground, he removed his sandals from beneath. (Exodus 3:5) And when Joshua conquered Jericho, he removed his sandals from beneath. (Joshua 5:15) And His people were commanded to remove any idols from their midst. (1 Samuel 7:3) And His people were commanded to remove any evil from His sight. (Isaiah 1:16) And we pray that the evil one is removed from beneath, that the Lord may rule from above. (Psalms 119:29)

And when they offered the sacrifice on the altar, He commanded them to remove the fat from their offering. (Leviticus 3:4) And the fat that was cut from the sacrifice of the altar, was burned up as smoke as a soothing aroma. (Leviticus 4:31) And when the red heifer burned on the temple, a priest was commanded to clean up the ashes. (Numbers 19:9) And the time will soon come when the sacrifice is man's flesh, and the fat from their hoarding shall be oil for the fire, and the blood of their wine shall burn into chaff. (Revelation 6:6)

Yet God can remove His wrath when He pleases. For when God turned His wrath onto Pharaoh, by Moses' request the plagues were withheld. (Exodus 8:8/ Exodus 10:17) And when the people would trust in His ways, the Lord would remove any illness from their midst. (Exodus 23:25) For when the serpents would poison His people, the Lord would remove the death of their sting, with the serpent that sat on the standard. (Numbers 21:6-9) And the Lord can remove all death from our lives, if we walk and obey in His ways. (Psalms 39:10)

"The serpent on the standard" Numbers 21:6-9

For the Lord commands man to turn from the flesh, and to "remove the foreskins of your heart". (Jeremiah 4:4) But man has removed his heart from the Lord, thus His "wrath shall go forth like fire." (Isaiah 29:13) And the vine shall be sheared from the earth, and the grapes shall be thrown in the press. (Isaiah 5:5) And their gold shall be thrown in the fire, and be smelted away with the lye. (Isaiah 1:25) And the turban shall be stripped from the beast of the earth, and the crown from the beast of the sea. (Ezekiel 21:26) And the sons and the prince of perdition, shall be stripped of their robes, and shall tremble in the starkness of their sins. (Ezekiel 26:16) "Therefore thus says the Lord, 'Behold, I am about to remove you from the face of the earth. This year you are going to die, because you have counseled rebellion against the Lord.'" (Jeremiah 28:16)

And the prophets and martyrs and saints, shall remove all the stones that lie in God's way. And the stone before Joshua is the seven that have returned. (Zechariah 3:9/Revelation 18:21) For the stone is "seven lamps of fire burning before the throne, which are the seven spirits of God". (Revelation 4:5) And the seven spirits of God are the "seven horns and seven eyes, which are the seven spirits of God, sent out into all the earth". (Revelation 5:6) And the seven spirits of God are "the seven angels who had the seven plagues coming out of the temple". (Revelation 15:6) And the seven angels of God carry seven bowls of wrath. "Then one of the four living creatures gave to the seven angels seven golden bowls full of the wrath of God". (Revelation 15:7) "And the temple was filled with smoke from the glory of God and from His power; and no one was able to enter the temple until the seven plagues of the seven angels were finished." (Revelation 15:8) And the remnant of His people have returned, to remove the abominations of the earth. (Ezekiel 11:18) For the prophets and martyrs and saints, shall remove all the stones that lie in His path, and shall build up a standard for the people. (Isaiah 62:10)

And when the final ruler of this earth is destroyed, the Lamb "will remove the heart of stone from your flesh and give you a heart of flesh." (Ezekiel 36:26) For the ruler that the Lamb shall destroy is called Death, and Death shall be swallowed for all time. Then men's tears shall be wept from their faces. (Revelation 21:4/Isaiah 25:8) And their tears will no longer be of sorrow and of pain. But their tears shall be of gladness and of joy. "Then you will call, and the Lord will answer; you will cry, and He will say, 'Here I am.'" (Isaiah 58:9)

XX

The Century of Service to the Lord (NASB)

To Andrew the Prophet

Completed February 4, 2008

Twenty is the year of service to the Temple

"These were the sons of Levi according to their fathers' households, even the heads of the fathers' households of those of them who were counted, in the number of names by their census , doing the work for the service of the house of the Lord, from **twenty** years old and upward. For David said, 'The Lord God of Israel has given rest to His people, and He dwells in Jerusalem forever. Also, the Levites will no longer need to carry the tabernacle and all its utensils for its service.' For by the last words of David the sons of Levi were numbered from **twenty** years old and upward" 2 Chronicles 23:24-27

"Now in the second year of their coming to the house of God at Jerusalem in the second month, Zerubbabel the son of Shealtiel and Jeshua the son of Jozadak and the rest of their brothers the priests and the Levites, and all who came from the captivity to Jerusalem, began the work and appointed the Levites from **twenty** years and older to oversee the work of the house of the Lord." Ezra 3:8

The tabernacle was built in twenty in semblance of the final kingdom

"You shall make the boards for the tabernacle: **twenty** boards for the south side and for the second side of the tabernacle, on the north side, **twenty** boards" Exodus 26:18, 20

"And its pillars shall be **twenty**, with their **twenty** sockets of bronze; the hooks of the pillars and their bands shall be of silver. Likewise for the north side in length there shall be hangings one hundred cubits long, and its **twenty** pillars with their **twenty** sockets of bronze; the hooks of the pillars and their bands shall be of silver." Exodus 27:10-11

The temple was built in twenty in semblance of the final kingdom

"Then he prepared an inner sanctuary within the house in order to place there the ark of the covenant of the Lord. The inner sanctuary was **twenty** cubits in length, twenty cubits in width, and **twenty** cubits in height, and he overlaid it with pure gold." 1 Kings 6:19-20

"Now he made the room of the holy of holies: its length across the width of the house was **twenty** cubits, and its width was **twenty** cubits; and he overlaid it with fine gold, amounting to 600 talents." 2 Chronicles 3:8

"The wingspan of the cherubim was **twenty** cubits; the wing of one, of five cubits, touched the wall of the house, and its other wing, of five cubits, touched the wing of the other cherub." 2 Chronicles 3:11

"Then he made a bronze altar, **twenty** cubits in length and **twenty** cubits in width and ten cubits in height." 2 Chronicles 4:1

"It came about at the end of **twenty** years in which Solomon had built the two houses, the house of the Lord and the king's house" 1 Kings 9:10

"He measured its length, **twenty** cubits, and the width, twenty cubits, before the nave; and he said to me, 'This is the most holy place.'" Ezekiel 41:4

Twenty was the age to go to war

"Everyone who is numbered, from **twenty** years old and over, shall give the contribution to the Lord." Exodus 30:14

"So all the numbered men of the sons of Israel by their fathers' households, from **twenty** years old and upward, whoever was able to go out to war in Israel, even all the numbered men were 603,550." Numbers 1:45-46

"Then it came about after the plague, that the Lord spoke to Moses and to Eleazar the son of Aaron the priest, saying, 'Take a census of all the congregation of the sons of Israel from **twenty** years old and upward, by their fathers' households, whoever is able to go out to war in Israel.' So Moses and Eleazar the priest spoke with them in the plains of Moab by the Jordan at Jericho, saying, 'Take a census of the people from **twenty** years old and upward, as the Lord has commanded Moses.'" Numbers 26:1-4

"For when they went up to the valley of Eshcol and saw the land, they discouraged the sons of Israel so that they did not go into the land which the Lord had given them. So the Lord's anger burned in that day, and He swore, saying, 'None of the men who came up from Egypt, from **twenty** years old and upward, shall see the land which I swore to Abraham, to Isaac and to Jacob; for they did not follow Me fully, except Caleb the son of Jephunneh the Kenizzite and Joshua the son of Nun, for they have followed the Lord fully.' So the Lord's anger burned against Israel, and He made them wander in the wilderness forty years, until the entire generation of those who had done evil in the sight of the Lord was destroyed." Numbers 32:9-13

Twenty is the final battle of Har-Magedon

"'Behold, I am coming like a thief. Blessed is the one who stays awake and keeps his clothes, so that he will not walk about naked and men will not see his shame.' And they gathered them together to the place which in Hebrew is called Har-Magedon." (Revelation 16:14-16)

Twenty is the Age of Service to the Lord

Twenty is the age of service to the Lord. For twenty was the age when the Levites were allowed, to "work for the service of the house of the Lord." (2 Chronicles 23:24, 27) And twenty was the age when the priests were allowed, "to oversee the work of the house of the Lord". (Ezra 3:8) And now the work of the house has been handed to His servants, to complete the works of the Lord.

For the tabernacle was built in twenty. For the tabernacle was held by twenty boards, on the north and south side of the Covenant. (Exodus 26:18, 20) And the tabernacle was held up by twenty pillars, inserted in twenty sockets of brass. (Exodus 27:10-11)

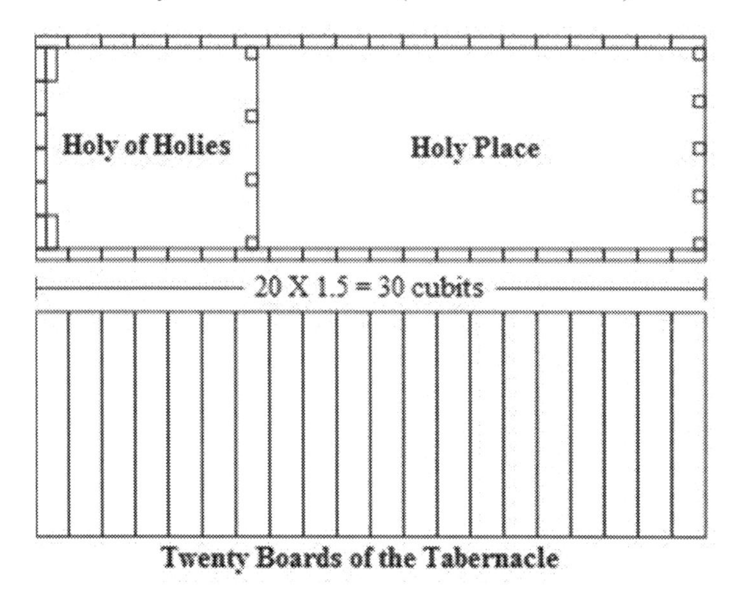

Twenty Boards of the Tabernacle

And the temple was built in twenty, in semblance of the kingdom to come. "Then he prepared an inner sanctuary within the house in order to place there the ark of the covenant of the Lord. The inner sanctuary was ***twenty*** cubits in length, ***twenty*** cubits in width, and ***twenty*** cubits in height, and he overlaid it with pure gold." (1 Kings 6:19-20/2 Chronicles 3:8) And the cherubim of the covenant was built in twenty. "The wingspan of the cherubim was ***twenty*** cubits; the wing of one, of five cubits, touched the wall of the house, and its other wing, of five cubits, touched the wing of the other cherub." (2 Chronicles 3:11) For the framework of the cherubim formed twenty (XX).

And the altar of bronze was in semblance of twenty, "***twenty*** cubits in length and ***twenty*** cubits in width." (2 Chronicles 4:1) For "in her was found the blood of prophets and of saints and of all who have been slain on the earth." (Revelation 18:24) For the prophet had foretold, that the true Temple would be twenty. "He measured its length, ***twenty*** cubits, and the width, ***twenty*** cubits, before the nave; and he said to me, 'This is the most holy place.'" (Ezekiel 41:4)

And twenty was the age to fight for His kingdom. For Moses was commanded by God, to take "a census of the people from twenty years old and upward". (Numbers 26:1-4) For the Lord commanded each man of twenty and upward, to give to the kingdom of God. (Exodus 30:14) And the men who were twenty and upward, were ordered to fight for the land. (Numbers 1:45-46) But when Israel marched into battle, the sons were great cowards in the land. "So the Lord's anger burned in that day, and He swore, saying, 'None of the men who came up from Egypt, from twenty years old and upward, shall see the land which I swore to Abraham, to Isaac and to Jacob; for they did not follow Me fully, except Caleb the son of Jephunneh the Kenizzite and Joshua the son of Nun, for they have followed the Lord fully.' So the Lord's anger burned against Israel, and He made them wander in the wilderness forty years, until the entire generation of those who had done evil in the sight of the Lord was destroyed." (Numbers 32:9-13)

And now that the end has drawn near, the war of Satan's servants shall begin. For in the year of twenty and twelve, the final battle on earth shall be fought. And those who have done evil in the sight of the Lord, shall gather and fight the last battle in shame. "'Behold, I am coming like a thief. Blessed is the one who stays awake and keeps his clothes, so that he will not walk about naked and men will not see his shame.' And they gathered them together to the place which in Hebrew is called Har-Magedon." (Revelation 16:14-15)

XXI

God Multiplies Our Product (NASB)

To Andrew the Prophet

Completed February 7, 2008

God multiplies His creation

"God blessed them, saying, 'Be fruitful and **multiply**, and fill the waters in the seas, and let birds **multiply** on the earth.'" Genesis 1:22

"God blessed them; and God said to them, 'Be fruitful and **multiply**, and fill the earth, and subdue it; and rule over the fish of the sea and over the birds of the sky and over every living thing that moves on the earth.'" Genesis 1:28

"And God blessed Noah and his sons and said to them, 'Be fruitful and **multiply**, and fill the earth.'" Genesis 9:1

"Now when Abram was ninety-nine years old, the Lord appeared to Abram and said to him,'I am God Almighty; walk before Me, and be blameless. I will establish My covenant between Me and you, and I will **multiply** you exceedingly.'" Genesis 17:1-2

"As for Ishmael, I have heard you; behold, I will bless him, and will make him fruitful and will **multiply** him exceedingly. He shall become the father of twelve princes, and I will make him a great nation." Genesis 17:20

"Then God appeared to Jacob again when he came from Paddan-aram, and He blessed him. God said to him, 'Your name is Jacob; You shall no longer be called Jacob, But Israel shall be your name.' Thus He called him Israel. God also said to him, 'I am God Almighty; be fruitful and **multiply**; a nation and a company of nations shall come from you, and kings shall come forth from you" Genesis 35:9-11

The Lord multiplies the blessings of those who follow Him

"All the commandments that I am commanding you today you shall be careful to do, that you may live and **multiply**, and go in and possess the land which the Lord swore to give to your forefathers." Deuteronomy 8:1

"Then it shall come about, because you listen to these judgments and keep and do them, that the Lord your God will keep with you His covenant and His lovingkindness which He swore to your forefathers. He will love you and bless you and **multiply** you; He will also bless the fruit of your womb and the fruit of your ground, your grain and your new wine and your oil, the increase of your herd and the young of your flock, in the land which He swore to your forefathers to give you. You shall be blessed above all peoples; there will be no male or female barren among you or among your cattle." Deuteronomy 7:12-14

His children multiply His Word

"'It shall be in those days when you are **multiplied** and increased in the land,' declares the Lord, 'they will no longer say, 'The ark of the covenant of the Lord.' And it will not come to mind, nor will they remember it, nor will they miss it, nor will it be made again.'" Jeremiah 3:16

"But the word of the Lord continued to grow and to be **multiplied**." Acts 12:24

"Now He who supplies seed to the sower and bread for food will supply and **multiply** your seed for sowing and increase the harvest of your righteousness." 2 Corinthians 9:10

God multiplies punishment on those who disobey Him

"To the woman He said, 'I will greatly **multiply** your pain in childbirth, in pain you will bring forth children; yet your desire will be for your husband, and he will rule over you.'" Genesis 3:16

"You shall speak all that I command you, and your brother Aaron shall speak to Pharaoh that he let the sons of Israel go out of his land. But I will harden Pharaoh's heart that I may **multiply** My signs and My wonders in the land of Egypt. When Pharaoh does not listen to you, then I will lay My hand on Egypt and bring out My hosts, My people the sons of Israel, from the land of Egypt by great judgments. The Egyptians shall know that I am the Lord, when I stretch out My hand on Egypt and bring out the sons of Israel from their midst." Exodus 7:2-5

"It shall come about that as the Lord delighted over you to prosper you, and **multiply** you, so the Lord will delight over you to make you perish and destroy you; and you will be torn from the land where you are entering to possess it." Deuteronomy 28:63

"The Lord has become like an enemy. He has swallowed up Israel; He has swallowed up all its palaces, He has destroyed its strongholds and **multiplied** in the daughter of Judah mourning and moaning." Lamentations 2:5

"Moreover, he did not humble himself before the Lord as his father Manasseh had done, but Amon **multiplied** guilt. Finally his servants conspired against him and put him to death in his own house." 2 Chronicles 33:23-24

"The sorrows of those who have bartered for another god will be **multiplied**; I shall not pour out their drink offerings of blood, nor will I take their names upon my lips." Psalms 16:4

Man has multiplied evil in the land

"And you have spoken arrogantly against Me and have **multiplied** your words against Me; I have heard it." Ezekiel 35:13

"You also played the harlot with the Egyptians, your lustful neighbors, and **multiplied** your harlotry to make Me angry. You also **multiplied** your harlotry with the land of merchants, Chaldea, yet even with this you were not satisfied." Ezekiel 16:26, 29

"You have **multiplied** your slain in this city, filling its streets with them." Ezekiel 11:6

"Yet the fool **multiplies** words. No man knows what will happen, and who can tell him what will come after him?" Ecclesiastes 10:14

And man shall pay for their multiple sins

"The more they **multiplied**, the more they sinned against Me; I will change their glory into shame." Hosea 4:7

"For Israel has forgotten his Maker and built palaces; and Judah has **multiplied** fortified cities, but I will send a fire on its cities that it may consume its palatial dwellings." Hosea 8:14

"'What are your **multiplied** sacrifices to Me?' says the Lord. 'I have had enough of burnt offerings of rams and the fat of fed cattle; and I take no pleasure in the blood of bulls, lambs or goats.'" Isaiah 1:11

"So when you spread out your hands in prayer, I will hide My eyes from you; yes, even though you **multiply** prayers, I will not listen. Your hands are covered with blood." Isaiah 1:15

And the remnant shall return

"Then I Myself will gather the remnant of My flock out of all the countries where I have driven them and bring them back to their pasture, and they will be fruitful and **multiply**." Jeremiah 23:3

"I will make a covenant of peace with them; it will be an everlasting covenant with them. And I will place them and **multiply** them, and will set My sanctuary in their midst forever." Ezekiel 37:26

"I will **multiply** the fruit of the tree and the **produce** of the field, so that you will not receive again the disgrace of famine among the nations." Ezekiel 36:30

The product of our works shall be tested by God

"With the fruit of a man's mouth his stomach will be satisfied; he will be satisfied with the **product** of his lips." Proverbs 18:20

"Give her the **product** of her hands, and let her works praise her in the gates." Proverbs 31:31

"You shall say to them, 'When you have offered from it the best of it, then the rest shall be reckoned to the Levites as the **product** of the threshing floor, and as the **product** of the wine vat.'" Numbers 18:30

"His winnowing fork is in His hand to thoroughly clear His threshing floor, and to gather the wheat into His barn; but He will burn up the chaff with unquenchable fire." Luke 3:17

"And the wine press was trodden outside the city, and blood came out from the wine press, up to the horses' bridles, for a distance of two hundred miles." Revelation 14:20

"He gave also their crops to the grasshopper and the **product** of their labor to the locust." Psalms 78:46

"Then out of the smoke came locusts upon the earth, and power was given them, as the scorpions of the earth have power.4 They were told not to hurt the grass of the earth, nor any green thing, nor any tree, but only the men who do not have the seal of God on their foreheads." Revelation 9:3-4

"Let the creditor seize all that he has, and let strangers plunder the **product** of his labor." Psalms 109:11

Our Produce Shall Be Multiplied in His Kingdom

Multiplication is a formidable power, for it is the power of adding multiple copies of one number. And God is a formidable Power, for by His great power His creation has multiplied. For when He made the creatures of the earth, He "blessed them, saying, 'Be fruitful and ***multiply***, and fill the waters in the seas, and let birds ***multiply*** on the earth.'" (Genesis 1:22) And when God created Adam and Eve, "God blessed them; and God said to them, 'Be fruitful and ***multiply***, and fill the earth, and subdue it; and rule over the fish of the sea and over the birds of the sky and over every living thing that moves on the earth.'" (Genesis 1:28) And when He destroyed mankind in the flood, He saved a remnant through Noah. And "God blessed Noah and his sons and said to them, 'Be fruitful and ***multiply***, and fill the earth.'" (Genesis 9:1) And God said to Abraham the father of all nations, "I am God Almighty; walk before Me, and be blameless. I will establish My covenant between Me and you, and I will ***multiply*** you exceedingly." (Genesis 17:1-2) And God blessed Ishmael and made Islam a great nation, "as for Ishmael, I have heard you; behold, I will bless him, and will make him fruitful and will ***multiply*** him exceedingly. He shall become the father of twelve princes, and I will make him a great nation." (Genesis 17:20) And God blessed Jacob and made Israel a great nation, "I am God, Almighty; be fruitful and ***multiply***; a nation and a company of nations shall come from you, and kings shall come forth from you." (Genesis 35:9-11)

And God multiplies the blessings, of those who obey His commands. For He promised that His people would multiply if they obeyed His commands. (Deuteronomy 7:12-14) For "all the commandments that I am commanding you today you shall be careful to do, that you may live and multiply, and go in and possess the land which the Lord swore to give to your forefathers." (Deuteronomy 8:1) And the prophecy foretold when mankind had multiplied, that the Son of God would return. For "it shall be in those days when you are multiplied and increased in the land, that the Son of God would come to establish a new covenant with mankind. And the covenant of old would not be missed, nor the temple be built of earthenware again." (Jeremiah 3:16)

And the signature of multiplication is a cross "x". And it was through the cross of Christ's crucifixion and death, that the "word of the Lord continued to grow and to be multiplied." (Acts 12:24) And in mathematics, the product of a series is represented by the symbol π. And the product of the series is:

$$\prod_{i=m}^{n} x_i = x_m \times x_{m+1} \times x_{m+2} \cdots x_{n-1} \times x_n$$

Where x is the defining equation, i is the variable, m is the first factor in a series, and n is the last factor in a series.

And as we know π represents the Son of God. And the Son continues to multiply our produce, for He "who supplies seed to the sower and bread for food will supply and multiply your seed for sowing and increase the harvest of your righteousness". (2 Corinthians 9:10)

But God also multiplies the suffering, of those who transgress His commands. For when Eve transgressed the Lord in the Garden, "to the woman He said, 'I will greatly multiply your pain in childbirth, in pain you will bring forth children.'" (Genesis 3:16) And when Pharaoh transgressed His people, God multiplied his suffering in the land. (Exodus 7:2-5) And when the people transgressed His commands, He cast them from the land they had possessed. (Deuteronomy 28:63) And He tore down their palaces and multiplied their mourning. (Lamentations 2:5) And the leaders who multiplied the guilt of the people, were punished and sentenced to death. (2 Chronicles 33:23-24) And those who worshipped false idols, God multiplied their suffering and death. (Psalms 16:4)

For "you have spoken arrogantly against Me and have multiplied your words against Me." (Ezekiel 35:13) For mankind has multiplied their harlotry and sins. (Ezekiel 16:26, 29) And mankind has multiplied their murderous deeds. (Ezekiel 11:6) Yet men play the fool and continue to sin. "Yet the fool multiplies words. No man knows what will happen, and who can tell him what will come after him?" (Ecclesiastes 10:14)

But now the judgment has come, when the Judge will change their glory into shame. For "the more they multiplied, the more they sinned against Me; I will change their glory into shame." (Hosea 4:7) For man has forgotten His Maker, but has built large palaces and fortified his cities. But now "I will send a fire on its cities that it may consume its palatial dwellings." (Hosea 8:14) And when that day comes, the multiplication of their prayers will be worthless. (Isaiah 1:11) "So when you spread out your hands in prayer, I will hide My eyes from you; yes, even though you multiply prayers, I will not listen. Your hands are covered with blood." (Isaiah 1:15)

But the remnant of His flock shall return to the Lord. (Jeremiah 23:3) And the Lord shall seal a covenant of peace. (Ezekiel 37:26) For His people will be free from the famine of nations, and will multiply the produce of their fields. (Ezekiel 36:30)

And as the outcome of multiplying is the product, the outcome of man's work is his *produce*. (Acts 12:24) For he who has multiplied God's Word, the produce of his work will satisfy him, and God "will be satisfied with the product of his lips." (Proverbs 18:20) And she who has served in the fear of the Lord, will receive the produce of her work, and her name will be praised at the gateway. (Proverbs 31:31) For the perfect shall be lifted, but the blemished shall be left, "as the product of the threshing floor, and as the product of the wine vat." (Numbers 18:30/Luke 3:17/Revelation 14:20) And the product of the threshing shall be given to the locusts. (Psalms 78:46/Revelation 9:3-4) And that which is taken, shall be given to His servants. (Psalms 101:11) For "I tell you that to everyone who has, more shall be given, but from the one who does not have, even what he does have shall be taken away." (Luke 19:26)

And once the wrath is complete, all shall partake of the produce of His kingdom. For if we take into account the product of the series π:

$$\prod_{i=m}^{n} x_i = \textbf{kingdom of heaven}$$

Π = product of the Son of God

x = work of the laborers

i = Spirit's intervention

m = first of the laborers

n = last of the laborers

"For the kingdom of heaven is like a landowner who went out early in the morning to hire laborers for his vineyard. When he had agreed with the laborers for a denarius for the day, he sent them into his vineyard. And he went out about the third hour and saw others standing idle in the market place; and to those he said, 'You also go into the vineyard, and whatever is right I will give you.' And so they went. Again he went out about the sixth and the ninth hour, and did the same thing. And about the eleventh hour he went out and found others standing around; and he said to them, 'Why have you been standing here idle all day long?' They said to him, 'Because no one hired us.' He said to them, 'You go into the vineyard too.' When evening came, the owner of the vineyard said to his foreman, 'Call the laborers and pay them their wages, beginning with the last group to the first.' When those hired about the eleventh hour came, each one received a denarius . When those hired first came, they thought that they would receive more; but each of them also received a denarius. When they received it, they grumbled at the landowner, saying, 'These last men have worked only one hour, and you have made them equal to us who have borne the burden and the scorching heat of the day.' But he answered and said to one of them, 'Friend, I am doing you no wrong; did you not agree with me for a denarius? 'Take what is yours and go, but I wish to give to this last man the same as to you. Is it not lawful for me to do what I wish with what is my own? Or is your eye envious because I am generous ?' So the last shall be first, and the first last." (Matthew 20:1-16)

XXII

Division Divides the People

To Andrew the Prophet

Completed February 9, 2008

God divided the waters from the beginning

"Now a river flowed out of Eden to water the garden; and from there it **divided** and became four rivers." Genesis 2:10

"As for you, lift up your staff and stretch out your hand over the sea and **divide** it, and the sons of Israel shall go through the midst of the sea on dry land." Exodus 14:16

"Elijah took his mantle and folded it together and struck the waters, and they were **divided** here and there, so that the two of them crossed over on dry ground." 2 Kings 2:8

"To Him who **divided** the Red Sea asunder, for His lovingkindness is everlasting." Psalms 136:13

God calls on us to divide our gifts with our neighbors

"Now if the household is too small for a lamb, then he and his neighbor nearest to his house are to take one according to the number of persons in them; according to what each man should eat, you are to **divide** the lamb." Exodus 12:4

"If one man's ox hurts another's so that it dies, then they shall sell the live ox and divide its price equally; and also they shall **divide** the dead ox." Exodus 21:35

"Is it not to **divide** your bread with the hungry and bring the homeless poor into the house; when you see the naked, to cover him; And not to hide yourself from your own flesh?" Isaiah 58:7

He commanded His people to divide the spoils

"They shall **divide** it into seven portions; Judah shall stay in its territory on the south, and the house of Joseph shall stay in their territory on the north." Joshua 18:5

"and said to them, "Return to your tents with great riches and with very much livestock, with silver, gold, bronze, iron, and with very many clothes; **divide** the spoil of your enemies with your brothers." Joshua 22:8

"It is better to be humble in spirit with the lowly than to **divide** the spoil with the proud." Proverbs 16:19

"The enemy said, 'I will pursue, I will overtake, I will **divide** the spoil; My desire shall be gratified against them; I will draw out my sword, my hand will destroy them.'" Exodus 15:9

And the Son has come to divide mankind

"Do you suppose that I came to grant peace on earth? I tell you, no, but rather **division**; for from now on five members in one household will be **divided**, three against two and two against three. They will be **divided**, father against son and son against father, mother against daughter and daughter against mother, mother-in-law against daughter-in-law and daughter-in-law against mother-in-law." Luke 12:51-53

"Others were saying, 'This is the Christ.' Still others were saying, 'Surely the Christ is not going to come from Galilee, is He? Has not the Scripture said that the Christ comes from the descendants of David, and from Bethlehem, the village where David was?' So a **division** occurred in the crowd because of Him." John 7:41-43

"And Jesus uttered a loud cry, and breathed His last. And the veil of the temple was torn in two from top to bottom." Mark 15:37-38

"Therefore, I will allot Him a portion with the great, and He will **divide** the booty with the strong; because He poured out Himself to death, and was numbered with the transgressors; yet He Himself bore the sin of many, and interceded for the transgressors." Isaiah 53:12

And the spoils shall be divided amongst His people

"He measured it on the four sides; it had a wall all around, the length five hundred and the width five hundred, to **divide** between the holy and the profane." Ezekiel 42:20

"Behold, a day is coming for the Lord when the spoil taken from you will be **divided** among you." Zechariah 14:1

"You shall multiply the nation, You shall increase their gladness; they will be glad in Your presence as with the gladness of harvest, as men rejoice when they **divide** the spoil." Isaiah 9:3

"Thus says the Lord God, 'This shall be the boundary by which you shall **divide** the land for an inheritance among the twelve tribes of Israel; Joseph shall have two portions.'" Ezekiel 47:13

"and I will make them one nation in the land, on the mountains of Israel; and one king will be king for all of them; and they will no longer be two nations and no longer be **divided** into two kingdoms." Ezekiel 37:22

The Son has the Power to Divide

"Any kingdom divided against itself is laid waste; and any city or house divided against itself will not stand." (Matthew 12:25)

In mathematics division is defined as an operation, whose function is the inverse of multiplication. And the Son has the power to multiply His kingdom. And the Son has the power to divide up the kingdom. And the signature of multiplication is a cross (X). And the signature of division is a line called a *vinculum* (/). And the factor above the vinculum is the numerator, and the factor below the vinculum is the denominator. And vinculum is the Latin for "bonds" and for "chains", for the kingdom is divided by bonds and by chains. For the numerator are the number who are bonded as His servants. For the time shall soon come "to reward Your ***bond***-servants the prophets and the saints and those who fear Your name, the small and the great." (Revelation 11:18) And the denominator are those who are chained by their sins. For they are "those who dwelt in darkness and in the shadow of death, prisoners in misery and ***chains***, because they had rebelled against the words of God and spurned the counsel of the Most High." (Psalms 107:10-11)

$$\frac{\textbf{Numerator}}{\textbf{Denominator}} = \frac{\textbf{His numbered servants}}{\textbf{Those demonized by the dominion of sin}}$$

For the power of the Most High has divided from the beginning. For from the Garden of Eden, the Creator divided the rivers. (Genesis 2:10) And from the Egyptian shore, the Lord divided the Red Sea. (Exodus 14:16) And from Elijah's mantle, the Lord divided the Jordan River. (2 Kings 2:8) And we give thanks for His lovingkindness, "to Him who divided the Red Sea asunder, for His lovingkindness is everlasting" (Psalms 136:13)

And the Lord commanded His people to divide up their gifts. For the Lord commanded His people to divide up the lamb. (Exodus 12:4) And His people were ordered to divide up their debts. (Exodus 21:35) And He ordered His people to divide up their bread: to feed the orphans and widows, to shelter the weak and the homeless, and to clothe the poor and the weary. (Isaiah 58:7)

And His people entered the land, and were commanded to divide up the land. (Joshua 18:5) And they were commanded to divide up the spoils. (Joshua 22:8) And His people were warned, "it is better to be humble in spirit with the lowly than to divide the spoil with the proud." (Proverbs 16:19) But His people were not humble but proud; and thus His words were fulfilled, "the enemy said, 'I will pursue, I will overtake, I will divide the spoil; My desire shall be gratified against them; I will draw out my sword, my hand will destroy them.'" (Exodus 15:9)

And the Son of God divides man against man. For the Son did not come to bring peace, but division and strife amongst men. For as He said, "Do you suppose that I came to grant peace on earth? I tell you, no, but rather division; for from now on five members in one household will be divided, three against two and two against three. They will be divided, father against son and son against father, mother against daughter and daughter against mother, mother-in-law against daughter-in-law and daughter-in-law against mother-in-law." (Luke 12:51-53) For division is derived from the Latin word *di* which means "two", and *vision* which means "see". For some have been given vision to see, but others have kept their eyes closed and remain blind. "But blessed are your eyes, because they see; and your ears, because they hear. For truly I say to you that many prophets and righteous men desired to see what you see, and did not see it, and to hear what you hear, and did not hear it." (Matthew 13:16-17) For many would see the works of His hands, yet divided they remained to the Truth. (John 7:41-43)

And through His death on the cross, the veil was divided into two. (Mark 15:37-38) And the day will soon come when the King shall return, and "He will divide the booty with the strong." (Isaiah 53:12) But the walls will divide the profane from the holy. (Ezekiel 42:20) For as the prophecy says "leave out the court which is outside the temple and do not measure it, for it has been given to the nations; and they will tread under foot the holy city for forty-two months." (Revelation 11:2) For the day will soon come when the spoil is divided. (Zechariah 14:1) And His servants will give praise in His presence, and they will divide the spoils of the earth. (Isaiah 9:3) For the spoils of the earth is the oil of man's sins, that shall burn in the fire of judgment. And the Lord shall divide the spoils with the nations. (Ezekiel 47:13) And He "will make them one nation in the land, on the mountains of Israel; and one king will be king for all of them; and they will no longer be two nations and no longer be divided into two kingdoms." (Ezekiel 37:22)

XXIII

God Holds the Scales of Equality (NASB)

To Andrew the Prophet

Completed February 11, 2008

God requires an equal balance

"A just **balance** and **scales** belong to the Lord; All the weights of the bag are His concern." Proverbs 16:11

"If a man seduces a virgin who is not engaged, and lies with her, he must pay a dowry for her to be his wife. If her father absolutely refuses to give her to him, he shall pay money **equal** to the dowry for virgins." Exodus 22:16-17

"Now if a Levite comes from any of your towns throughout Israel where he resides, and comes whenever he desires to the place which the Lord chooses, then he shall serve in the name of the Lord his God, like all his fellow Levites who stand there before the Lord. They shall eat **equal** portions, except what they receive from the sale of their fathers' estates." Deuteronomy 18:8

"If a fellow countryman of yours becomes so poor he has to sell part of his property, then his nearest kinsman is to come and buy back what his relative has sold. Or in case a man has no kinsman, but so recovers his means as to find sufficient for its redemption, then he shall calculate the years since its sale and refund the **balance** to the man to whom he sold it, and so return to his property." Leviticus 25:25-28

"For this is not for the ease of others and for your affliction, but by way of **equality—at** this present time your abundance being a supply for their need, so that their abundance also may become a supply for your need, that there may be **equality.**" 2 Corinthians 8:13-14

God despises a false balance

"A false **balance** is an abomination to the Lord, but a just weight is His delight." Proverbs 11:1

"Can I justify wicked **scales** and a bag of deceptive weights?" Micah 6:11

"Hear this, you who trample the needy, to do away with the humble of the land, saying, 'When will the new moon be over, so that we may sell grain, and the Sabbath, that we may open the wheat market, to make the bushel smaller and the shekel bigger, and to cheat with dishonest **scales**, so as to buy the helpless for money and the needy for a pair of sandals, and that we may sell the refuse of the wheat?' The Lord has sworn by the pride of Jacob, 'Indeed, I will never forget any of their deeds because of this will not the land quake and everyone who dwells in it mourn? Indeed, all of it will rise up like the Nile, and it will be tossed about and subside like the Nile of Egypt.'" Amos 8:4-8

But most of all, He despises those who believe they hold the scales of balance

"'To whom then will you liken Me that I would be his **equal**?' says the Holy One." Isaiah 40:25

"Who, although He existed in the form of God, did not regard **equality** with God a thing to be grasped." Philippians 2:6

"You turn things around! Shall the potter be considered as **equal** with the clay, that what is made would say to its maker, 'He did not make me'; Or what is formed say to him who formed it, 'He has no understanding'?" Isaiah 29:16

"It even magnified itself to be **equal** with the Commander of the host; and it removed the regular sacrifice from Him, and the place of His sanctuary was thrown down." Daniel 8:11

The treasures of God shall not be balanced on earth, but shall be brought up to Heaven

"Do not store up for yourselves **treasures** on earth, where moth and rust destroy, and where thieves break in and steal. But store up for yourselves **treasures** in heaven, where neither moth nor rust destroys, and where thieves do not break in or steal; for where your **treasure** is, there your heart will be also." Matthew 6:19-21

"Therefore it says, 'When He ascended on high, He led captive a host of captives, And He gave **gifts** to men.'" Ephesians 4:8

"From the standpoint of the gospel they are enemies for your sake, but from the standpoint of God's choice they are beloved for the sake of the fathers; for the **gifts** and the calling of God are irrevocable." Romans 11:28-29

But the Debts of the Earth Shall be Balanced

"Who has measured the waters in the hollow of His hand, And marked off the heavens by the span, and calculated the dust of the earth by the measure, And weighed the mountains in a **balance** And the hills in a pair of **scales**? Behold, the nations are like a drop from a bucket, and are regarded as a speck of dust on the **scales**; behold, He lifts up the islands like fine dust." Isaiah 40:12, 15

"When He broke the third seal, I heard the third living creature saying, 'Come.' I looked, and behold, a black horse; and he who sat on it had a pair of **scales** in his hand." Revelations 6:5

When the Third Seal is Broken the Balance Shall Begin (11:11)

On September 11, 2001 the Trade Towers fell, and the beast from the sea was released. (Revelations 13:1-10) And the scales which balance this earth, were weighed down by the sons of perdition. But woe to these oppressors, for "a false balance is an abomination to the Lord." (Proverbs 11:1)

In the infancy of mathematics, the symbol for equality was "ς " for "aequalis", which in Latin means "equal" or "balanced". And "a just balance and scales belong to the Lord;

all the weights of the bag are His concern." (Proverbs 16:11) And to simplify the mathematical equation, the symbol of aequalis became two parallel lines | | . And in numerology "11" is the symbol for justice. "For the Lord loves justice and does not forsake His godly ones; they are preserved forever." (Psalms 37:28) And the mathematician Recorde laid out the equal sign as "=". For He said the equal sign is a "pair of parallels of one length, because no two things can be more equal". (*The Whetstone of Witte* - Robert Recorde, 1557) And there are no two things that are more equal, that than the equality of His heavenly Kingdom. "The city is laid out as a square, and its length is as great as the width; and he measured the city with the rod, fifteen hundred miles; its length and width and height are equal." (Revelation 21:16)

And even in the days of antiquity, when their daughters were given away, God told them to equal the balance. (Exodus 22:16-17) And the priests who tended God's temple, were commanded to equal their portions. (Deuteronomy 18:8) And God commanded His people, to equal their land and possessions. (Leviticus 25:25-28) And the Lord commanded His disciples, to equal the portions of their gifts. (2 Corinthians 8:13-14)

But man in his greed and deception, has weighed down the scale with false weights. But "a false weight is an abomination to the Lord." (Proverbs 1:11) And by weighing down the scales with greed and injustice, the Lord would warn man "can I justify wicked scales and a bag of deceptive weights?" (Micah 6:11) For man has withheld the food from the widows, and pillaged the poor with the scale of deception. Thus "the Lord has sworn by the pride of Jacob, 'Indeed, I will never forget any of their deeds because of this will not the land quake and everyone who dwells in it mourn?'" (Amos 8:4-8)

And our Lord most despises the sons of perdition, for they believe they control the weights of His scale. "'To whom then will you liken Me that I would be his equal?' says the Holy One." (Isaiah 40:25) For even the Son who "existed in the form of God, did not regard equality with God a thing to be grasped". (Philippians 2:6) Yet the sons of perdition have made themselves judge, and boast to the Maker who has formed them: "He has no understanding".

(Isaiah 29:16) And the son of perdition shall even magnify "itself to be equal with the Commander of the host", and shall remove the witness from Him, and shall throw down the place of His sanctuary. (Daniel 8:11)

But when the scales are balanced, His servants shall be given great treasures. For the treasures and gifts they are given, shall not be balanced in the fire. For as the Son had taught them, "store up for yourselves treasures in heaven, where neither moth nor rust destroys, and where thieves do not break in or steal; for where your treasure is, there your heart will be also." (Matthew 6:19-21) For when He died on the cross, and "when He ascended on high, He led captive a host of captives, and He gave gifts to men.'" (Ephesians 4:8) ***Thus the spirit of His servants shall live forever***, "for the gifts and the calling of God are irrevocable." (Romans 11:28-29)

But the scales have been falsely weighted, by the deception and wickedness of men. Yet it is God "who has measured the waters in the hollow of His hand, and marked off the heavens by the span, and calculated the dust of the earth by the measure, and weighed the mountains in a balance and the hills in a pair of scales. Behold, the nations are like a drop from a bucket, and are regarded as a speck of dust on the scales; behold, He lifts up the islands like fine dust." (Isaiah 40:12, 15) "For we know Him who said, 'Vengeance is Mine, I will repay.' And again, 'The Lord will judge His people.' It is a terrifying thing to fall into the hands of the living God." (Hebrews 10:30-31) And in July of 2009, the third seal shall be opened. For Uganda shall be consumed in a great famine, and the Judge shall return to balance His scales. For as it was foretold "when He broke the third seal, I heard the third living creature saying, 'Come.' I looked, and behold, a black horse; and he who sat on it had a pair of scales in his hand." (Revelations 6:5)

The Balance of God's Hands

40

Satan's Rule Shall Soon End (NASB)

To Andrew the Prophet

Completed February 20, 2008

Forty was necessary to clean man from their transgressions

"For after seven more days, I will send rain on the earth **forty** days and **forty** nights; and I will blot out from the face of the land every living thing that I have made." Genesis 7:4

"So the Lord's anger burned against Israel, and He made them wander in the wilderness **forty** years, until the entire generation of those who had done evil in the sight of the Lord was destroyed." Numbers 32:13

"Now the sons of Israel again did evil in the sight of the Lord, so that the Lord gave them into the hands of the Philistines **forty** years." Judges 13:1

"A man's foot will not pass through it, and the foot of a beast will not pass through it, and it will not be inhabited for **forty** years. So I will make the land of Egypt a desolation in the midst of desolated lands. And her cities, in the midst of cities that are laid waste, will be desolate **forty** years; and I will scatter the Egyptians among the nations and disperse them among the lands. At the end of **forty** years I will gather the Egyptians from the peoples among whom they were scattered." Ezekiel 29:11-13

But God provided for His people during these forty years

"The sons of Israel ate the manna **forty** years, until they came to an inhabited land; they ate the manna until they came to the border of the land of Canaan." Exodus 16:35

"For the Lord your God has blessed you in all that you have done; He has known your wanderings through this great wilderness. These

forty years the Lord your God has been with you; you have not lacked a thing." Deuteronomy 2:7

"You shall remember all the way which the Lord your God has led you in the wilderness these **forty** years, that He might humble you, testing you, to know what was in your heart, whether you would keep His commandments or not." Deuteronomy 8:2

His servants would fast for forty days

"Moses entered the midst of the cloud as he went up to the mountain; and Moses was on the mountain **forty** days and **forty** nights." Exodus 24:18

"So he was there with the Lord **forty** days and forty nights; he did not eat bread or drink water. And he wrote on the tablets the words of the covenant, the Ten Commandments." Exodus 34:28

"I fell down before the Lord, as at the first, **forty** days and nights; I neither ate bread nor drank water, because of all your sin which you had committed in doing what was evil in the sight of the Lord to provoke Him to anger. So I fell down before the Lord the **forty** days and nights, which I did because the Lord had said He would destroy you." Deuteronomy 9:18, 25

"I, moreover, stayed on the mountain **forty** days and forty nights like the first time, and the Lord listened to me that time also; the Lord was not willing to destroy you." Deuteronomy 10:10

"So he arose and ate and drank, and went in the strength of that food **forty** days and **forty** nights to Horeb, the mountain of God." 1 Kings 19:8

Christ prepared His disciples in forty days

"And He was in the wilderness **forty** days being tempted by Satan; and He was with the wild beasts, and the angels were ministering to Him." Mark 1:13

"To these He also presented Himself alive after His suffering, by many convincing proofs, appearing to them over a period of **forty** days and speaking of the things concerning the kingdom of God." Acts 1:13

And one out of one-thousand would enter His kingdom

"For they are a nation lacking in counsel, and there is no understanding in them. Would that they were wise, that they understood this, that they would discern their future! How could one chase a **thousand**, and two put ten **thousand** to flight, unless their Rock had sold them, and the Lord had given them up?" Deuteronomy 32:28-30

"One of your men puts to flight a **thousand**, for the Lord your God is He who fights for you, just as He promised you." Joshua 23:10

"These of the sons of Gad were captains of the army; he who was least was equal to a hundred and the greatest to a **thousand**." 1 Chronicles 12:14

"He found a fresh jawbone of a donkey, so he reached out and took it and killed a **thousand** men with it." Judges 15:15

"A **thousand** may fall at your side and ten **thousand** at your right hand, but it shall not approach you." Psalms 91:7

"One **thousand** will flee at the threat of one man; you will flee at the threat of five, until you are left as a flag on a mountain top and as a signal on a hill." Isaiah 30:17

"Do you not yet understand or remember the five loaves of the five **thousand**, and how many baskets full you picked up?" Matthew 16:9

"May the Lord, the God of your fathers, increase you a **thousand-fold** more than you are and bless you, just as He has promised you!" Deuteronomy 1:11

"For a day in Your courts is better than a **thousand** outside. I would rather stand at the threshold of the house of my God than dwell in the tents of wickedness." Psalms 84:10

"For a **thousand** years in Your sight are like yesterday when it passes by, or as a watch in the night." Psalms 90:4

"But do not let this one fact escape your notice, beloved, that with the Lord one day is like a **thousand** years, and a **thousand** years like one day." 2 Peter 3:8

"Then I saw thrones, and they sat on them, and judgment was given to them. And I saw the souls of those who had been beheaded because of their testimony of Jesus and because of the word of God, and those who had not worshiped the beast or his image, and had not received the mark on their forehead and on their hand; and they came to life and reigned with Christ for a **thousand** years. The rest of the dead did not come to life until the **thousand** years were completed. This is the first resurrection. Blessed and holy is the one who has a part in the first resurrection; over these the second death has no power, but they will be priests of God and of Christ and will reign with Him for a **thousand** years. When the **thousand** years are completed, Satan will be released from his prison" Revelation 20:4-7

The Son of Man

The rule of this earth has been given to Satan, but he has been released for just a short time. (Revelation 20:3) And His people have yearned for the Son to return. For the wait of His people has been in forty. For they wandered through the desert for forty long years, for doing grave deeds in His sight. (Numbers 32:13) And the Philistines ruled for forty long years, when His people had worshiped false idols. (Judges 13:1) And the king of Egypt was warned by the Lord, "behold, I am against you, Pharaoh king of Egypt, the great monster that lies in the midst of his rivers, that has said, 'My Nile is mine, and I myself have made it.'" (Ezekiel 29:3) Thus for forty long years his kingdom was destroyed, and the land laid in ruin and waste. (Ezekiel 29:11-13)

But though they had wandered for forty long years, the Lord gave them food that His people may live. For His people had manna from heaven, when they came to the land of their fathers. (Exodus 16:35) And for forty years God lived in their midst, thus they lacked not one single thing. (Deuteronomy 2:7) And He said to His people, "you shall remember all the way which the Lord your God has led you in the wilderness these forty years, that He might humble you, testing you, to know what was in your heart, whether you would keep His commandments or not." (Deuteronomy 8:2)

And by forty His servant would fast day and night, that God would have mercy on the people. For Moses would fast forty days and forty nights, when God gave the people His commands. (Exodus 24:18, 34:28) For "I fell down before the Lord the forty days and nights, which I did because the Lord had said He would destroy you." (Deuteronomy 9:18,25) And when God returned the broken covenant, Moses fasted forty days and forty nights. (Deuteronomy 10:10) And Elijah fasted forty days and forty nights, so that God withheld His wrath from the people. (1 Kings 19:8)

And the Father sent His Son down to earth for mankind. And He fasted in the wilderness forty days and forty nights, to prepare for the ministry that would come. (Mark 1:13) And the Son of Man taught us the Word, and the Son of Man poured out His blood, and the Son of Man arose from the dead. And after He had arisen, the Son of Man returned for forty days and forty nights, to prepare His disciples for the Spirit and the ministry that would come. And through the power of the Spirit, they would spread the works and His Word throughout the earth. (Acts 1:13) And after forty days with His servants, the Son of Man would ascend into heaven. And He has patiently awaited for the Day that shall soon come.

And the number for forty is the Greek letter Mu (**M**), from the Phoenician word for water which is "**M**", and the Egyptian glyph for water which is "**N**". And as the Lord warned Noah, "I will send rain on the earth forty days and forty nights; and I will blot out from the face of the land every living thing that I have made." (Genesis 7:4) But "I establish My covenant with you; and all flesh shall never again be cut off by the **water** of the flood, neither shall there again be a flood to destroy the earth."(Genesis 9:11)

The Letters M and N for Water

Egyptian Phoenician Greek Roman

For God shall not destroy man by water again. But this time all flesh shall be destroyed by the fire, and the fire shall consume all the earth.

And if we recall the Greek alphabet:

α β γ Δ ϵ ζ Η θ ι Κ λ Μ Ν ξ Ο Π ρ σ τ υ φ Χ Υ Ω

And as one can see, **M and N** cross through the middle of the alphabet. And the Son was there at the beginning, and the **Son of MaN crossed in the middle**, and the Son shall rise at the end. For the Son of **MaN** is the narrow gate to the Father. For the Son of Man taught us, "enter through the narrow gate; for the gate is wide and the way is broad that leads to destruction, and there are many who enter through it. For the gate is small and the way is narrow that leads to life, and there are few who find it." (Matthew 7:13-14)

And the Roman number **M** represents one thousand. And due to the transgressions of His people, one thousand would flee by the chase of just one. (Deuteronomy 32:28-30) But those who obeyed His commands, their power was greater than one thousand. (Joshua 23:10/1 Chronicles 12:14/Judges 15:15/Psalms 91:7/Isaiah 30:17) And as He would teach His disciples, He would multiply five loaves to five thousand. (Matthew 16:9) And the steadfastness of just one disciple, would yield one thousand more servants. For M comes from the Latin word *mille* for one thousand, as a mile is *mille passus* for "one thousand paces". And He taught His disciples to go one thousand paces, for "whoever forces you to go one mile, go with him two."(Matthew 5:41) And His yield shall be great for "He will increase you a thousand-fold more than you are and bless you, just as He has promised you!" (Deuteronomy 1:11) And truly "a day in

Your courts is better than a thousand outside. I would rather stand at the threshold of the house of my God than dwell in the tents of wickedness." (Psalms 84:10)

And did He not promise that this Day would soon come? For "with the Lord one day is like a thousand years, and a thousand years like one day." (2 Peter 3:8/Psalms 90:4) For the Son of Man rose into heaven, and the Son reigned for one thousand years. And Satan was bound for one thousand years, but the ruler has since returned to the earth. For the "Holy" Crusades" began in one thousand AD, and the "church" would defile and destroy the Holy Land. But one thousand was all the earth's ruler was given, for his sons have been cast down to earth. And all men shall soon live, for the rule of evil shall soon end. (Revelation 13:7-13/Revelation 20:4-7)

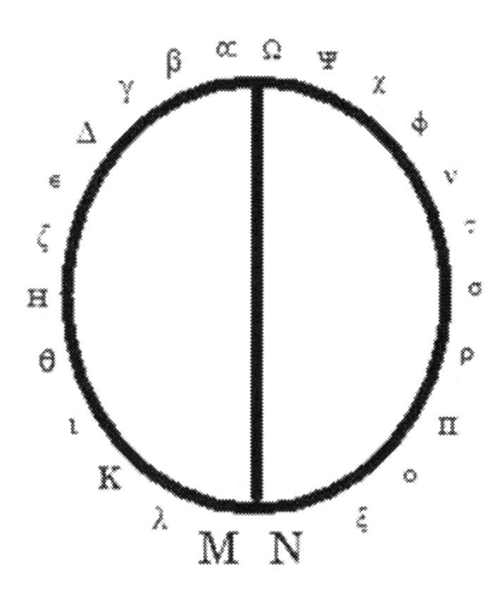

Son of Man

α the Beginning

"In the beginning was the Word, and the Word was with God, and the Word was God. He was in the beginning with God. All things came into being through Him, and apart from Him nothing came into being that has come into being. In Him was life, and the life was the Light of men. The Light shines in the darkness, and the darkness did not comprehend it." (John 1:1-5)

MaN the Middle

"If anyone wishes to come after Me, he must deny himself, and take up his cross and follow Me. For whoever wishes to save his life will lose it; but whoever loses his life for My sake will find it. For what will it profit a man if he gains the whole world and forfeits his soul? Or what will a man give in exchange for his soul? For the ***Son of Man*** is going to come in the glory of His Father with His angels, and will then repay every man according to his deeds. Truly I say to you, there are some of those who are standing here who will not taste death until they see the ***Son of Man*** coming in His kingdom." (Matthew 16:24-28)

Then I turned to see the voice that was speaking with me. And having turned I saw seven golden lampstands; and in the middle of the lampstands I saw one like a ***son of man***, clothed in a robe reaching to the feet, and girded across His chest with a golden sash." (Revelation 1:12-13)

Ω the End

"It is done. I am the Alpha and the Omega, the beginning and the end. I will give to the one who thirsts from the spring of the water of life without cost." (Revelation 21:6)

100

One Hundred Shall Enter His Court (NASB)

To Andrew the Prophet

Completed February 22, 2008

His kingdom has increased one hundred fold

"Now Abraham was one **hundred** years old when his son Isaac was born to him." Genesis 21:5

"Now Isaac sowed in that land and reaped in the same year a **hundredfold**. And the Lord blessed him, and the man became rich, and continued to grow richer until he became very wealthy; for he had possessions of flocks and herds and a great household, so that the Philistines envied him" Genesis 26:12-14

"Joab said, 'May the Lord add to His people a **hundred** times as many as they are! But, my lord the king, are they not all my lord's servants? Why does my lord seek this thing? Why should he be a cause of guilt to Israel?'" 1 Chronicles 21:3

"But Joab said to the king, 'Now may the Lord your God add to the people a **hundred** times as many as they are, while the eyes of my lord the king still see; but why does my lord the king delight in this thing?'" 2 Samuel 24:3

Their military strength increased one hundred fold

"Five of you will chase a **hundred**, and a **hundred** of you will chase ten thousand, and your enemies will fall before you by the sword." Leviticus 26:8

"So Gideon and the **hundred** men who were with him came to the outskirts of the camp at the beginning of the middle watch, when they had just posted the watch; and they blew the trumpets and smashed the pitchers that were in their hands." Judge 7:19

"These of the sons of Gad were captains of the army; he who was least was equal to a **hundred** and the greatest to a thousand." 1 Chronicles 12:14

"He hired also **100,000** valiant warriors out of Israel for one **hundred** talents of silver." 2 Chronicles 25:6

"Amaziah said to the man of God, 'But what shall we do for the **hundred** talents which I have given to the troops of Israel?' And the man of God answered, 'The Lord has much more to give you than this.'" 2 Chronicles 25:9

"And when Jesus entered Capernaum, a **centurion** came to Him, imploring Him, and saying, 'Lord, my servant is lying paralyzed at home, fearfully tormented.' Jesus said to him, 'I will come and heal him.' But the **centurion** said, 'Lord, I am not worthy for You to come under my roof, but just say the word, and my servant will be healed. For I also am a man under authority, with soldiers under me; and I say to this one, 'Go!' and he goes, and to another, 'Come!' and he comes, and to my slave, 'Do this!' and he does it.' Now when Jesus heard this, He marveled and said to those who were following, 'Truly I say to you, I have not found such great faith with anyone in Israel. I say to you that many will come from east and west, and recline at the table with Abraham, Isaac and Jacob in the kingdom of heaven; but the sons of the kingdom will be cast out into the outer darkness; in that place there will be weeping and gnashing of teeth.' And Jesus said to the **centurion**, 'Go; it shall be done for you as you have believed.' And the servant was healed that very moment." Matthew 8:5-13

One hundred treasures in this world is worth nothing

"A rebuke goes deeper into one who has understanding than a **hundred** blows into a fool." Proverbs 17:10

"If a man fathers a **hundred** children and lives many years, however many they be, but his soul is not satisfied with good things and he does not even have a proper burial, then I say, "Better the miscarriage than he." Ecclesiastes 6:3

"Although a sinner does evil a **hundred** times and may lengthen his life, still I know that it will be well for those who fear God, who fear Him openly." Ecclesiastes 8:12

"For thus says the Lord God, "The city which goes forth a thousand strong Will have a **hundred** left, And the one which goes forth a **hundred** strong Will have ten left to the house of Israel." Amos 5:3

His tabernacle was built in one hundred

"You shall make the court of the tabernacle. On the south side there shall be hangings for the court of fine twisted linen one **hundred** cubits long for one side." Exodus 27:9

"Likewise for the north side in length there shall be hangings one **hundred** cubits long, and its twenty pillars with their twenty sockets of bronze; the hooks of the pillars and their bands shall be of silver." Exodus 27:11

"The length of the court shall be one **hundred** cubits, and the width fifty throughout, and the height five cubits of fine twisted linen, and their sockets of bronze." Exodus 27:18

"The **hundred** talents of silver were for casting the sockets of the sanctuary and the sockets of the veil; one **hundred** sockets for the **hundred** talents, a talent for a socket." Exodus 38:27

His temple was built in one hundred

"He made chains in the inner sanctuary and placed them on the tops of the pillars; and he made one **hundred** pomegranates and placed them on the chains." 2 Chronicles 3:16

"There were ninety-six exposed pomegranates; all the pomegranates numbered a **hundred** on the network all around." Jeremiah 52:23

"He also made ten tables and placed them in the temple, five on the right side and five on the left. And he made one **hundred** golden bowls." 2 Chronicles 4:8

"Then Eliashib the high priest arose with his brothers the priests and built the Sheep Gate; they consecrated it and hung its doors. They consecrated the wall to the Tower of the **Hundred** and the Tower of Hananel." Nehemiah 3:1

"And above the Gate of Ephraim, by the Old Gate, by the Fish Gate, the Tower of Hananel and the Tower of the **Hundred**, as far as the Sheep Gate; and they stopped at the Gate of the Guard." Nehemiah 12:39

The final temple is built in one hundred

"Then he measured the width from the front of the lower gate to the front of the exterior of the inner court, a **hundred** cubits on the east and on the north." Ezekiel 40:19

"The inner court had a gate opposite the gate on the north as well as the gate on the east; and he measured a **hundred** cubits from gate to gate." Ezekiel 40:23

"The inner court had a gate toward the south; and he measured from gate to gate toward the south, a **hundred** cubits." Ezekiel 40:27

"He measured the court, a perfect square, a **hundred** cubits long and a **hundred** cubits wide; and the altar was in front of the temple." Ezekiel 40:47

"Then he measured the temple, a **hundred** cubits long; the separate area with the building and its walls were also a **hundred** cubits long. Also the width of the front of the temple and that of the separate areas along the east side totaled a **hundred** cubits. He measured the length of the building along the front of the separate area behind it, with a gallery on each side, a **hundred** cubits; he also measured the inner nave and the porches of the court." Ezekiel 41:13-15

"Along the length, which was a **hundred** cubits, was the north door; the width was fifty cubits. Before the chambers was an inner walk ten cubits wide, a way of one **hundred** cubits; and their openings were on the north. For the length of the chambers which were in the outer court was fifty cubits; and behold, the length of those facing the temple was a **hundred** cubits." Ezekiel 42:2, 4, 8

Our rewards in heaven are increased one-hundred fold

"But that he will receive a **hundred** times as much now in the present age, houses and brothers and sisters and mothers and children and farms, along with persecutions; and in the age to come, eternal life." Mark 10:30

"'Other seed fell into the good soil, and grew up, and produced a crop a **hundred** times as great.' As He said these things, He would call out, 'He who has ears to hear, let him hear.'" Luke 8:8

"And the one on whom seed was sown on the good soil, this is the man who hears the word and understands it; who indeed bears fruit and brings forth, some a **hundredfold**, some sixty, and some thirty." Matthew 13:23

One Hundred Shall Enter His Temple

The number one hundred is symbolized by the Roman numeral C, for *centum* which means "to multiply ten by ten times". And ten by ten symbolizes the hands of our God, for He multiplies mankind by the product of His hands.

C = X x X = 100

C = 100 = Product of God's Hands

And through the faith of our forefathers, God multiplied mankind by one hundredfold. For Abraham was one hundred years old, when Isaac was born to the people. (Genesis 21:5) And when Isaac was given the land, "Isaac sowed in that land and reaped in the same year a hundredfold." (Genesis 26:12-14) And when David took a census of the people, they had increased by one hundredfold. (1 Chronicles 21:3/2 Samuel 24:3)

And the Lord multiplies the strength of His people, and their power by one hundredfold. (Leviticus 26:8) For a centurion commands an army of one hundred men, as the Holy One commands all of creation. And it was by the strength of God's hands, that Gideon conquered the Midianites with just one hundred men. (Judge 7:19) And through the power of God's hands, the army of David would increase one hundredfold. (1 Chronicles 12:14) And when Amaziah battled the Syrians, he purchased 100,000 soldiers with one hundred talents. But God instructed Amaziah to fight by God's power, so the centum of soldiers was released, yet the strength of his men increased one hundredfold. (2 Chronicles 25:9) And it was by a centurion's faith in Christ's power, that the power of God's strength was revealed. (Matthew 8:5-13)

Yet despite the power of God's strength, the heart of Amaziah would turn to idolatry, and the anger of God would destroy him. For one hundred treasures on earth are worth nothing in heaven. And "a rebuke goes deeper into one who has understanding than a hundred blows into a fool." (Proverbs 17:10) And even a hundred children without the rest of one's soul is worth nothing. (Ecclesiastes 6:3) For on that day the Lord's wrath will burn, and "the city which goes forth a thousand strong will have a hundred left, and the one which goes forth a hundred strong will have ten left to the house of Israel." (Amos 5:3)

And the tabernacle was made in one hundred, in the image of the Temple of God. For the length of the tabernacle was one hundred cubits. (Exodus 27:9, 11, 18) And the sockets of the tabernacle were one hundred, and the cost of the sockets was one hundred talents. (Exodus 38:27) And the temple of God was made in one hundred, in the image of His kingdom's great produce. For the sanctuary contained the fruits of one hundred. (2 Chronicles 3:16/Jeremiah 52:23) And the fruits were placed in bowls of one hundred. (2 Chronicles 4:8) And the temple was guarded by the Tower of One Hundred. (Nehemiah 3:1)

And the number for one hundred is the Greek letter rho (ρ), which comes from the Egyptian glyphic for "head". And the Temple of God was built in one hundred, for the Son who is the head of His kingdom. And the Temple and the house shall measure one hundred. (Ezekiel 40:47/Ezekiel 42:2,4,8) For the Temple shall be given to the Groom, and the house to the people who worship the

Lord. But the court shall be left for the nations to destroy. "Get up and measure the temple of God and the altar, and those who worship in it. Leave out the court which is outside the temple and do not measure it, for it has been given to the nations; and they will tread under foot the holy city for forty-two months." (Revelation 11:1-2/ Ezekiel 40:19,23,27)

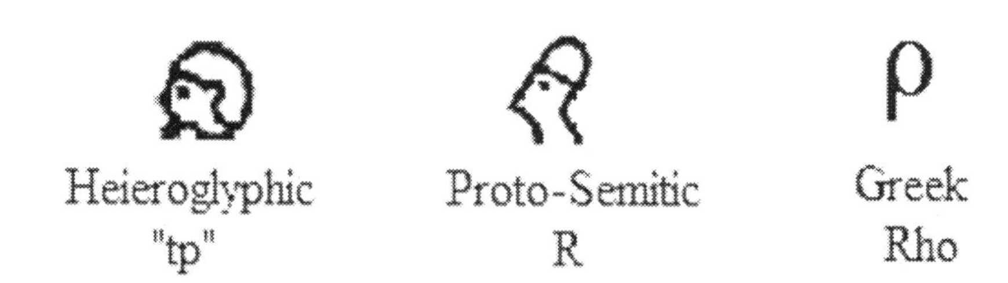

For the Lord has promised that the gifts of His children would increase by one hundredfold. "He will receive a hundred times as much now in the present age and in the age to come, eternal life." (Mark 10:30) And the good Seed has "produced a crop a hundred times as great." (Luke 8:8) For the fruit of the Seed has grown one hundredfold. "And those are the ones on whom seed was sown on the good soil; and they hear the word and accept it and bear fruit, thirty, sixty, and a hundredfold." (Mark 4:20)

But for those who have forsaken the Lord, He forewarned, "truly I say to you, you will not come out of there until you have paid up the last ***cent***." (Matthew 5:26) And the symbol of C represents Celsius degrees, and water will boil at one hundred degrees Celsius. And the anger of God has increased one hundredfold. Thus He will "rend the heavens and come down, that the mountains might quake at Your presence—as fire kindles the brushwood, ***as fire causes water to boil*** — to make Your name known to Your adversaries, that the nations may tremble at Your presence!" (Isaiah 64:1-2)

The Fire = 100°C

144,000

His Elect (NASB)

To Andrew the Prophet

Completed February 24, 2008

"And then the sign of the Son of Man will appear in the sky, and then all the tribes of the earth will mourn, and they will see the Son of Man coming on the clouds of the sky and great glory. And He will send forth His angels with a great trumpet and they will gather together **His elect** from the four winds, from one end of the sky to the other." Matthew 24:30-31

The 144,000

"And I heard the number of those who were sealed, **one hundred and forty-four thousand** sealed from every tribe of the sons of Israel: From the tribe of Judah, twelve thousand were sealed, from the tribe of Reuben twelve thousand, from the tribe of Gad twelve thousand, from the tribe of Asher twelve thousand, from the tribe of Naphtali twelve thousand, from the tribe of Manasseh twelve thousand, from the tribe of Simeon twelve thousand, from the tribe of Levi twelve thousand, from the tribe of Issachar twelve thousand, from the tribe of Zebulun twelve thousand, from the tribe of Joseph twelve thousand, from the tribe of Benjamin, twelve thousand were sealed." Revelation 7:4-8

"Then I looked, and behold, the Lamb was standing on **Mount Zion**, and with Him **one hundred and forty-four thousand**, having His name and the name of His Father written on their foreheads. And I heard a voice from heaven, like the sound of many waters and like the sound of loud thunder, and the voice which I heard was like the sound of harpists playing on their harps. And they sang a new song before the throne and before the **four living creatures** and the elders; and no one could learn the song except the **one hundred and forty-four thousand** who had been purchased from the earth. These are the ones who have not been defiled with women, for they have kept themselves chaste. These are the ones who follow the Lamb wherever He goes. These have been purchased from among men as **first fruits** to God and to the Lamb. And no lie was found in their mouth; they are blameless." Revelation 14:1-5

"It was also about these men that Enoch, in the seventh generation from Adam, prophesied, saying, 'Behold, the Lord came with many **thousands of His holy ones**, to execute judgment upon all, and to convict all the ungodly of all their ungodly deeds which they have done in an ungodly way, and of all the harsh things which ungodly sinners have spoken against Him." Jude 1:14-15

The prophets and martyrs and saints

"Thus says the Lord of hosts, 'Let your hands be strong, you who are listening in these days to these words from the mouth of the **prophets**, those who spoke in the day that the **foundation** of the house of the Lord of hosts was laid, to the end that the **temple** might be built.'" Zechariah 8:9

"I kept looking, and that horn was waging war with the **saints** and overpowering them until the Ancient of Days came and judgment was passed in favor of the **saints** of the Highest One, and the time arrived when the **saints** took possession of the kingdom." Daniel 7:21-22

"We give You thanks, O Lord God, the Almighty, who are and who were, because You have taken Your great power and have begun to reign. And the nations were enraged, and Your wrath came, and the time came for the dead to be judged, and the time to reward Your bond-servants the **prophets** and the **saints** and those who fear Your name, the small and the great, and to destroy those who destroy the earth." Revelation 11:17-18

"Righteous are You, who are and who were, O Holy One, because You judged these things; for they poured out the blood of **saints** and **prophets**, and You have given them blood to drink. They deserve it." Revelation 16:5-6

"Then a strong angel took up a stone like a great millstone and threw it into the sea, saying, 'So will Babylon, the great city, be thrown down with violence, and will not be found any longer. And the sound of harpists and musicians and flute-players and trumpeters will not be heard in you any longer; and no craftsman of any craft will be found in you any longer; and the sound of a mill will not be heard in you any longer; and the light of a lamp will not shine in you any longer; and the voice of the bridegroom and **bride** will not be heard in you any longer; for your merchants were the great men of the earth, because all the nations were deceived by your sorcery. And in her was found the blood of **prophets** and of **saints** and of all who have been slain on the earth." Revelation 18:21-24

"Let us rejoice and be glad and give the glory to Him, for the marriage of the Lamb has come and His **bride** has made herself ready. It was given to her to clothe herself in fine linen, bright and clean; for the fine linen is the righteous acts of the **saints.**" Revelation 19:7-8

The first fruits

"You shall bring the very first of the **first fruits** of your soil into the house of the Lord your God." Exodus 34:26

"As an offering of **first fruits** you shall bring them to the Lord, but they shall not ascend for a soothing aroma on the altar." Leviticus 2:12-13

"These are the ones who follow the Lamb wherever He goes. These have been purchased from among men as **first fruits** to God and to the Lamb. And no lie was found in their mouth; they are blameless." Revelation 14:4-5

"And not only this, but also we ourselves, having the **first fruits** of the Spirit, even we ourselves groan within ourselves, waiting eagerly for our adoption as sons, the redemption of our body." Romans 8:23

"For as in Adam all die, so also in Christ all will be made alive. But each in his own order: Christ the **first fruits**, after that those who are Christ's at His coming, then comes the end, when He hands over the kingdom to the God and Father, when He has abolished all rule and all authority and power." 1 Corinthians 15:22-24

The remnant of Israel

"For out of Jerusalem will go forth a **remnant**, and out of Mount **Zion** survivors. The zeal of the Lord will perform this." 2 Kings 19:31

"Now in that day the **remnant** of Israel, and those of the house of Jacob who have escaped, will never again rely on the one who struck them, but will truly rely on the Lord, the Holy One of Israel. A **remnant** will return, the **remnant** of Jacob, to the mighty God. For though your people, O Israel, may be like the sand of the sea, Only a **remnant** within them will return; a destruction is determined, overflowing with righteousness. For a complete destruction, one that is decreed, the Lord God of hosts will execute in the midst of the whole land." Isaiah 10:20-23

"Then it will happen on that day that the Lord Will again recover the second time with His hand

The **remnant** of His people, who will remain, From Assyria, Egypt, Pathros, Cush, Elam, Shinar, Hamath, And from the islands of the sea. And He will lift up a standard for the nations

And assemble the banished ones of Israel, And will gather the dispersed of Judah From the four corners of the earth." Isaiah 11:11-12

"The surviving **remnant** of the **house of Judah** will again take root downward and bear fruit upward. For out of Jerusalem will go forth a **remnant** and out of Mount **Zion** survivors. The zeal of the Lord of hosts will perform this." Isaiah 37:31-32

"Thus says the Lord of hosts, 'They will thoroughly glean as the vine the **remnant** of Israel;

Pass your hand again like a grape gatherer over the branches.'" Jeremiah 6:9

"Then I Myself will gather the **remnant** of My flock out of all the countries where I have driven them and bring them back to their pasture, and they will be fruitful and multiply. I will also raise up shepherds over them and they will tend them; and they will not be afraid any longer, nor be terrified, nor will any be missing,' declares the Lord." Jeremiah 23:3-4

"I will surely assemble all of you, Jacob, I will surely gather the **remnant** of Israel. I will put them together like sheep in the fold; like a flock in the midst of its pasture" Micah 2:12

"But I will leave among you A humble and lowly people, and they will take refuge in the name of the Lord. The **remnant** of Israel will do no wrong and tell no lies, nor will a deceitful tongue be found in their mouths; for they will feed and lie down with no one to make them tremble. Shout for joy, **O daughter of Zion**! Shout in triumph, O Israel! Rejoice and exult with all your heart, O daughter of Jerusalem!" Zephaniah 3:12-14

"'But now I will not treat the **remnant** of this people as in the former days,' declares the Lord of hosts. For there will be peace for the seed: the vine will yield its fruit, the land will yield its produce and the heavens will give their dew; and I will cause the **remnant** of this people to inherit all these things. It will come about that just as you were a curse among the nations, **O house of Judah** and house of Israel, so I will save you that you may become a blessing. Do not fear; let your hands be strong." Zechariah 8:11-13

The bride of the Bridegroom

"'Lift up your eyes and look around; All of them gather together, they come to you. As I live,' declares the Lord, 'You will surely put on all of them as jewels and bind them on as a **bride**'" Isaiah 49:18

"Then their offspring will be known among the nations, and their descendants in the midst of the peoples. All who see them will recognize them because they are the offspring whom the Lord has blessed. I will rejoice greatly in the Lord, My soul will exult in my God; for He has clothed me with garments of salvation, He has wrapped me with a robe of righteousness, as a bridegroom decks himself with a garland, and as a **bride** adorns herself with her jewels. For as the earth brings forth its sprouts, and as a garden causes the things sown in it to spring up, so the Lord God will cause righteousness and praise to spring up before all the nations." Isaiah 61:9-11

"For **Zion's** sake I will not keep silent, and for Jerusalem's sake I will not keep quiet, Until her righteousness goes forth like brightness, and her salvation like a torch that is burning. The nations will see your righteousness, and all kings your glory; and you will be called by a new name which the mouth of the Lord will designate. You will also be a crown of beauty in the hand of the Lord, and a royal diadem in the hand of your God. It will no longer be said to you, 'Forsaken ', nor to your land will it any longer be said, 'Desolate'; but you will be called, 'My delight is in her,' and your land, 'Married ; for the Lord delights in you, and to Him your land will be married. For as a young man marries a virgin, so your sons will marry you; and as the bridegroom rejoices over the **bride**, so your God will rejoice over you.'" Isaiah 62:1-5

"Thus says the Lord, 'Yet again there will be heard in this place, of which you say, 'It is a waste, without man and without beast,' that is, in the cities of Judah and in the streets of Jerusalem that are desolate, without man and without inhabitant and without beast, the voice of joy and the voice of gladness, the voice of the bridegroom and the voice of the **bride**, the voice of those who say, 'Give thanks to the Lord of hosts, for the Lord is good, for His lovingkindness is everlasting'; and of those who bring a thank offering into the house of the Lord. For I will restore the fortunes of the land as they were at first,' says the Lord." Jeremiah 33:10-11

"He who has the **bride** is the bridegroom; but the friend of the bridegroom, who stands and hears him, rejoices greatly because of the bridegroom's voice. So this joy of mine has been made full." John 3:29-30

"And the light of a lamp will not shine in you any longer; and the voice of the bridegroom and **bride** will not be heard in you any longer; for your merchants were the great men of the earth, because all the nations were deceived by your sorcery." Revelation 18:23

"The Spirit and the **bride** say, 'Come.' And let the one who hears say, 'Come.' And let the one who is thirsty come; let the one who wishes take the water of life without cost." Revelation 22:17

The tribe of Judah

"**Judah**, your brothers shall praise you; your hand shall be on the neck of your enemies; your father's sons shall bow down to you. **Judah** is a lion's whelp; from the prey, my son, you have gone up. He crouches, he lies down as a lion, And as a lion, who dares rouse him up? The scepter shall not depart from **Judah**, nor the ruler's staff from between his feet, until Shiloh comes, and to him shall be the obedience of the peoples." Genesis 49:8-10

"Now the lot for the **tribe** of the sons **of Judah** according to their families reached the border of Edom, southward to the wilderness of **Zin** at the extreme south." Joshua 15:1

"It came about when all Israel heard that Jeroboam had returned, that they sent and called him to the assembly and made him king over all Israel. None but the **tribe of Judah** followed the house of David." 1 Kings 12:20

"So the Lord was very angry with Israel and removed them from His sight; none was left except the **tribe of Judah**." 2 Kings 17:18

"But chose the **tribe of Judah**, Mount **Zion** which He loved." Psalms 78:68 31

"**Judah** became His sanctuary, Israel, His dominion." Psalms 114:2

"'Behold, the days are coming,' declares the Lord, 'When I will raise up for David a righteous Branch ; and He will reign as king and act wisely and do justice and righteousness in the land.

In His days **Judah** will be saved, and Israel will dwell securely; and this is His name by which He will be called, 'The Lord our righteousness.'" Jeremiah 23:5-6

"'Behold, days are coming,' declares the Lord, 'when I will fulfill the good word which I have spoken concerning the house of Israel and the house of Judah. In those days and at that time I will cause a righteous Branch of David to spring forth; and He shall execute justice and righteousness on the earth. In those days **Judah** will be saved and Jerusalem will dwell in safety; and this is the name by which she will be called: the Lord is our righteousness.'" Jeremiah 33:14-16

"'Behold, days are coming,' declares the Lord, 'when I will make a new covenant with the house of Israel and with the **house of Judah**'" Jeremiah 31:31

"I will restore the fortunes of **Judah** and the fortunes of Israel and will rebuild them as they were at first." Jeremiah 33:7-8

"For behold, in those days and at that time, when I restore the fortunes of **Judah** and Jerusalem, I will gather all the nations and bring them down to the valley of Jehoshaphat." Joel 3:1

"And in that day the mountains will drip with sweet wine, and the hills will flow with milk, and all the brooks of **Judah** will flow with water; and a spring will go out from the house of the Lord To water the valley of Shittim. Egypt will become a waste, and Edom will become a desolate wilderness, because of the violence done to the sons of **Judah**, in whose land they have shed innocent blood. But **Judah** will be inhabited forever and Jerusalem for all generations. And I will avenge their blood which I have not avenged, for the Lord dwells in **Zion**." Joel 3:18-21

"I will strengthen the **house of Judah**, and I will save the house of Joseph, and I will bring them back, because I have had compassion on them; and they will be as though I had not rejected them, for I am the Lord their God and I will answer them." Zechariah 10:6

"In that day I will make the clans of Judah like a firepot among pieces of wood and a flaming torch among sheaves, so they will consume on the right hand and on the left all the surrounding peoples, while the inhabitants of Jerusalem again dwell on their own sites in Jerusalem. The Lord also will save the tents of Judah first, so that the glory of the house of David and the glory of the inhabitants of Jerusalem will not be magnified above **Judah**." Zechariah 12:6-7

"'Behold, I am going to send My messenger, and he will clear the way before Me. And the Lord, whom you seek, will suddenly come to His temple; and the messenger of the covenant, in whom you delight, behold, He is coming,' says the Lord of hosts. But who can endure the day of His coming? And who can stand when He appears? For He is like a refiner's fire and like fullers soap. He will sit as a smelter and purifier of silver, and He will purify the sons of Levi and refine them like gold and silver, so that they may present to the Lord offerings in righteousness. Then the offering of **Judah** and Jerusalem will be pleasing to the Lord as in the days of old and as in former years." Malachi 3:1-4

"Gathering together all the chief priests and scribes of the people, he inquired of them where the Messiah was to be born. They said to him, 'In Bethlehem of Judea; for this is what has been written by the prophet: And you, Bethlehem of **Judah** means least among the leaders of **Judah**;

For out of you shall come forth a ruler will shepherd My people Israel" Matthew 2:4-6

"For it is evident that our Lord was descended from **Judah, a tribe** with reference to which Moses spoke nothing concerning priests" Hebrews 7:14

"Stop weeping; behold, the Lion that is from the **tribe of Judah**, the Root of David, has overcome so as to open the book and its seven seals." Revelation 5:5

The creature with four heads

"As I looked, behold, a storm wind was coming from the north, a great cloud with fire flashing forth continually and a bright light around it, and in its midst something like glowing metal in the midst of the fire. Within it there were figures resembling **four living beings.** And this was their appearance: they had human form. Each of them had four faces and four wings. Their legs were straight and their feet were like a calf's hoof, and they gleamed like burnished bronze. Under their wings on their four sides were human hands. As for the faces and wings of the four of them, their wings touched one another; their faces did not turn when they moved, each went straight forward. As for the form of their faces, each had the face of a man; all four had the face of a lion on the right and the face of a bull on the left, and all four had the face of an eagle. Such were their faces. Their wings were spread out above; each had two touching another being, and two covering their bodies. And each went straight forward; wherever the spirit was about to go, they would go, without turning as they went. In the midst of the living beings there was something that looked like burning coals of fire, like torches darting back and forth among the living beings. The fire was bright, and lightning was flashing from the fire. And the living beings ran to and fro like bolts of lightning." Ezekiel 1:4-14

"Out from the throne come flashes of lightning and sounds and peals of thunder. And there were seven lamps of fire burning before the throne, which are the seven Spirits of God; and before the throne there was something like a sea of glass, like crystal; and in the center and around the throne, **four living creatures** full of eyes in front and behind. The first creature was like a lion, and the second creature like a calf, and the third creature had a face like that of a man, and the fourth creature was like a flying eagle. And the four living creatures, each one of them having six wings, are full of eyes around and within; and day and night they do not cease to say, 'Holy, holy, holy is the Lord God, the Almighty, who was and who is and who is to come.'" Revelation 4:5-8

The daughter of Zion

"Let Mount **Zion** be glad, let the **daughters of Judah** rejoice because of Your judgments. Walk about **Zion** and go around her; count her towers; consider her ramparts; go through her palaces, that you may tell it to the next generation." Psalms 48:11-13

"Awake, awake, clothe yourself in your strength, O **Zion**; clothe yourself in your beautiful garments, O Jerusalem, the holy city; for the uncircumcised and the unclean will no longer come into you.

Shake yourself from the dust, rise up, O captive Jerusalem; loose yourself from the chains around your neck, O captive **daughter of Zion**." Isaiah 52:1-2

"Behold, the Lord has proclaimed to the end of the earth, say to the **daughter of Zion**, 'Lo, your salvation comes; behold His reward is with Him, and His recompense before Him.'" Isaiah 62:11

"As for you, tower of the flock, hill of the **daughter of Zion**, to you it will come—even the former dominion will come, the kingdom of the daughter of Jerusalem. Writhe and labor to give birth, **Daughter of Zion**, like a woman in childbirth; for now you will go out of the city, Dwell in the field, and go to Babylon. There you will be rescued; there the Lord will redeem you from the hand of your enemies. Arise and thresh, **daughter of Zion**, for your horn I will make iron and your hoofs I will make bronze, that you may pulverize many peoples, that you may devote to the Lord their unjust gain and their wealth to the Lord of all the earth." Micah 4:8 10, 13

"'Sing for joy and be glad, O **daughter of Zion**; for behold I am coming and I will dwell in your midst,' declares the Lord. Many nations will join themselves to the Lord in that day and will become My people. Then I will dwell in your midst, and you will know that the Lord of hosts has sent Me to you." Zechariah 2:10-11

"Say to the **daughter of Zion**, 'Behold your king is coming to you, gentle, and mounted on a donkey, even on a colt, the foal of a beast of burden.'" Matthew 21:5

The foundation of the Temple of God

"Then there was given me a measuring rod like a staff; and someone said, 'Get up and measure the **temple** of God and the altar, and those who worship in it.'" Revelation 11:1

"Now the altar hearth shall be **twelve** cubits long by **twelve** wide, square in its four sides."

Ezekiel 43:16

"Now the cherubim were standing on the right side of the **temple** when the man entered, and the cloud filled the inner court. Then the glory of the Lord went up from the cherub to the threshold of the **temple**, and the **temple** was filled with the cloud and the court was filled with the brightness of the glory of the Lord." Ezekiel 10:3-4

"Thus says the Lord of hosts, 'Behold, a man whose name is Branch, for He will branch out from where He is; and He will build the **temple** of the Lord. Yes, it is He who will build the temple of the Lord, and He who will bear the honor and sit and rule on His throne. Thus, He will be a priest on His throne, and the counsel of peace will be between the two offices. Now the crown will become a reminder in the **temple** of the Lord to Helem, Tobijah, Jedaiah and Hen the son of Zephaniah. Those who are far off will come and build the **temple** of the Lord. Then you will know that the Lord of hosts has sent me to you. And it will take place if you completely obey the Lord your God." Zechariah 6:12-15

"Do you not know that you are a **temple** of God and that the Spirit of God dwells in you? If any man destroys the **temple** of God, God will destroy him, for the **temple** of God is holy, and that is what you are." 1 Corinthians 3:16-17

"Or what agreement has the **temple** of God with idols? For we are the **temple** of the living God; just as God said, 'I will dwell in them and walk among them; and I will be their God, and they shall be my people.'" 2 Corinthians 6:16

"So then you are no longer strangers and aliens, but you are fellow citizens with the saints, and are of God's household, having been built on the **foundation** of the apostles and prophets, Christ Jesus Himself being the corner stone, in whom the whole building, being fitted together, is growing into a holy **temple** in the Lord, in whom you also are being built together into a dwelling of God in the Spirit" Ephesians 2:19-22

"He who overcomes, I will make him a pillar in the **temple** of My God, and he will not go out from it anymore; and I will write on him the name of My God, and the name of the city of My God, the new

Jerusalem, which comes down out of heaven from My God, and My new name." Revelation 3:12

"After these things I looked, and the **temple** of the tabernacle of testimony in heaven was opened, and the seven angels who had the seven plagues came out of the temple, clothed in linen, clean and bright, and girded around their chests with golden sashes. Then one of the four living creatures gave to the seven angels seven golden bowls full of the wrath of God, who lives forever and ever. And the **temple** was filled with smoke from the glory of God and from His power; and no one was able to enter the **temple** until the seven plagues of the seven angels were finished." Revelation 15:5-8

"I saw no **temple** in it, for the Lord God the Almighty and the Lamb are its **temple**. And the city has no need of the sun or of the moon to shine on it, for the glory of God has illumined it, and its lamp is the Lamb." Revelation 21:22-23

The True Temple of God

The Elect

The 144,000

His holy ones

The prophets and martyrs and saints

The first fruits

The remnant of Israel

The bride of the Bridegroom

The tribe of Judah

The creature with four heads

The daughter of Zion

The foundation of the Temple of God

The 144,000: The True Temple of God

On that day, the Son of Man shall rise in the clouds, the heavenly hosts shall give praises of thanksgiving, and the Elect of mankind shall be raised. For the archangels shall travel from the east to the west, from the north to the south, to the corners of the earth, and on chariots raise His Elect to the heavens. For the heavenly host "waits eagerly for the revealing of the sons of God." (Romans 8:19)

144 thousand **=** thousands of **hhl** (His holy ones)

For His Elect are one hundred and forty-four thousand, His holy ones slain for the kingdom of God. And those that are sealed are one hundred and forty-four thousand, for these are the twelve tribes by twelve thousand. (Revelation 7:4-8) And these are the bride of the Bridegroom, and these are the daughters of Zion. And "these have been purchased from among men as first fruits to God and to the Lamb. And no lie was found in their mouth; they are blameless." (Revelation 14:1-5) And behold the Lord comes with the 144 thousand. For "the first will be last and the last will be first", for the Lord comes with many thousands of His holy ones (hh1). "It was also about these men that Enoch, in the seventh generation from Adam, prophesied, saying, 'Behold, **the Lord came with many thousands of His holy ones**, to execute judgment upon all, and to convict all the ungodly of all their ungodly deeds which they have done in an ungodly way, and of all the harsh things which ungodly sinners have spoken against Him." (Jude 1:14-15)

And His Elect are the prophets and martyrs and saints, for they have served for the Lord since "the day that the foundation of the house of the Lord of hosts was laid, to the end that the temple might be built." (Zechariah 8:9) But the beast shall make war with His saints, until "the Ancient of Days comes, and judgment is passed in favor of the saints of the Highest One, and the time arrives when the saints take possession of the kingdom." (Daniel 7:21-22) For "the time has come for the dead to be judged, and the time to reward Your bond-servants the prophets and the saints and those who fear Your name, the small and the great, and to destroy those who destroy the earth." (Revelation 11:17-18) "For they poured out the blood of saints and prophets, and You have given them blood to drink. They deserve it." (Revelation 16:5-6) And the earth and its ruler shall be slain, and time and death shall seize to be. (Revelation 18:21-24) "For the marriage of the Lamb has come and His bride has made herself ready. It was given to her to clothe herself in fine linen, bright and clean; for the fine linen is the righteous acts of the saints." (Revelation 19:7-8)

And His Elect are the firstfruits of the harvest. For the people were told to save the firstfruits of their harvest. (Exodus 34:26) And the firstfruits were saved from the fire, for they were perfect and no fault could be found in them. (Leviticus 2:12-13) And His firstfruits will be saved from the fire, for "these have been purchased from among men as first fruits to God and to the Lamb. And no lie was found in their mouth; they are blameless." (Revelation 14:4-5) For we "having the first fruits of the Spirit, even we ourselves groan within ourselves, waiting eagerly for our adoption as sons, the redemption of our body." (Romans 8:23) But after the fire of judgment, all men shall be reconciled to the Father, "but each in his own order: Christ the first fruits, after that those who are Christ's at His coming, then comes the end, when He hands over the kingdom to the God and Father, when He has abolished all rule and all authority and power." (1 Corinthians 15:22-24)

And His Elect are the remnant of Israel. "For out of Jerusalem will go forth a remnant, and out of Mount Zion survivors. The zeal of the Lord will perform this." (2 Kings 19:31/Isaiah 37:31-32) But "though your people, O Israel, may be like the sand of the sea, only a remnant within them will return; a destruction is determined, overflowing with righteousness. For a complete destruction, one that is decreed, the Lord God of hosts will execute in the midst of the whole land." (Isaiah 10:20-23) "Then it will happen on that day that the Lord will again recover the ***second time*** with His hand the remnant of His people... and He will lift up a standard for the nations and assemble the banished ones of Israel, and will gather the dispersed of Judah from the four corners of the earth." (Isaiah

11:11-12) For the Son of Man "will thoroughly glean as the vine the remnant of Israel; pass your hand again like a grape gatherer over the branches." (Jeremiah 6:9) And He "will gather the remnant of My flock out of all the countries where I have driven them and bring them back to their pasture, and they will be fruitful and multiply. I will also raise up shepherds over them and they will tend them; and they will not be afraid any longer, nor be terrified, nor will any be missing." (Jeremiah 23:3-4/Micah 2:12) For "the remnant of Israel will do no wrong and tell no lies, nor will a deceitful tongue be found in their mouths." (Zephaniah 3:12-14) "For there will be peace for the seed: the vine will yield its fruit, the land will yield its produce and the heavens will give their dew; and I will cause the remnant of this people to inherit all these things." (Zechariah 8:11-13)

And His Elect are the bride of the Bridegroom. For "all of them gather together, they come to you. As I live You will surely put on all of them as jewels and bind them on as a bride'" (Isaiah 49:18) For they shall gather from all of the nations, and they shall be clothed in righteousness, "as a bridegroom decks himself with a garland, and as a bride adorns herself with her jewels." (Isaiah 61:9-11) And her "righteousness goes forth like brightness, and her salvation like a torch that is burning. And as the bridegroom rejoices over the bride, so your God will rejoice over you." (Isaiah 62:1-5) And the voice of His bride shall give praise, to God for His goodness and faithfulness. (Jeremiah 33:10-11) For His witness "rejoices greatly because of the bridegroom's voice. So this joy of mine has been made full." (John 3:29-30) But on that day, the bride shall be raised from the earth, and the fire of judgment shall burn forth. (Revelation 18:23) And when the last ruler has been slain, "the Spirit and the bride will say, 'Come.' And let the one who hears say, 'Come.' And let the one who is thirsty come; let the one who wishes take the water of life without cost." (Revelation 22:17)

And His Elect are the true tribe of Judah. For from the tribe of Judah, the Son would bear the scepter and the staff. (Genesis 49:8-10) And the land of Judah would touch the foothold of Zion. (Joshua 15:1) For the only tribe to follow the Lord was Judah. (1 Kings 12:20/2 Kings 17:18) Thus He "chose the tribe of Judah, Mount

Zion which He loved." (Psalms 78:68 31) And "Judah became His sanctuary, Israel, His dominion." (Psalms 114:2) And the prophets foretold that the Branch would return, and would spread His Word throughout the land. (Jeremiah 23:5-6/Jeremiah 33:14-16) And the Branch made a covenant with Judah (Jeremiah 31:31), to rebuild and restore them as at first. (Jeremiah 33:7-8/Joel 3:1) And the Branch shall avenge all their suffering. (Joel 3:18-21) "For I am the Lord their God and I will answer them." (Zechariah 10:6) For "the glory of the inhabitants of Jerusalem will not be magnified above Judah." (Zechariah 12:6-7/Malachi 3:1-4) Thus the Son of Judah would be born in Bethlehem. (Matthew 2:4-6) And the Lamb of Judah would be sacrificed in Jerusalem. (Hebrew 7:14) And the Lion of Judah shall return in the clouds. So "stop weeping; behold, the Lion that is from the tribe of Judah, the Root of David, has overcome so as to open the book and its seven seals." (Revelation 5:5)

And His Elect is the creature with four heads. (Ezekiel 1:4-14/ Revelation 4:5-8) For they were made in the image of God. And the first is the head of a man, for they were made in the image of the Father. And the second is the head of the sacrifice, for they were made in the image of the Lamb. And the third is the head of an eagle, for they were made in the image of the Spirit. And the fourth is the head of a lion, for they were made in the image of the Lion. "And day and night they do not cease to say, 'Holy, holy, holy is the Lord God, the Almighty, who was and who is and who is to come.'" (Revelation 4:8)

And His Elect is the daughter of Zion, and they "rejoice because of Your judgments." (Psalms 48:11-13) For when the Lamb entered Jerusalem, He said "to the daughter of Zion, 'Behold your king is coming to you, gentle, and mounted on a donkey, even on a colt, the foal of a beast of burden.'" (Matthew 21:5) And the bride shall be clothed in righteousness, when the chains are removed from her neck. (Isaiah 52:1-2) And the Lord will "say to the daughter of Zion, 'Lo, your salvation comes; behold His reward is with Him, and His recompense before Him.'" (Isaiah 62:11) And the daughter shall be saved from her enemies, and injustice shall be paid through His hands. (Micah 4:8,10, 13) And when the last debt has been paid, "many nations will join themselves to the Lord in that day and will become My people. Then I will dwell in your midst, and you will know that the Lord of hosts has sent Me to you." (Zechariah 2:10-11) And when the Lion returns in the clouds, "the first shall be last and the last shall be first". For when the daughter of Zion returns, the world will know of her sacrifice (NO 12).

ZION = NO 12 **The 12 tribes By 12 thousand**

And His Elect are the foundation of the Temple, for their sacrifice on the altar was in twelve. For the Temple of God has been measured, (Revelation 11:1) and the Temple is twelve by twelve thousand. (Revelation 7:4-8) And the altar is "twelve cubits long by twelve wide, square in its four sides." (Ezekiel 43:16) And the Temple will be filled with the glory of God. (Ezekiel 10:3-4) For the Son is the Branch of the Temple, and "He will branch out from where He is; and He will build the temple of the Lord." For they are the branches of the Vine. (Zechariah 6:12-15) And they are the foundation of the Temple. (1 Corinthians 3:16-17) For "just as God said, 'I will dwell in them and walk among them; and I will be their God, and they shall be my people.'" (2 Corinthians 6:16) For the Temple has "been built on the foundation of the apostles and prophets, Christ Jesus Himself being the corner stone." (Ephesians 2:19-22) And His martyrs are "a pillar in the temple of My God". (Revelation 3:12) And when the Son comes in the clouds, His Elect shall be the Temple, "and the temple was filled with smoke from the glory of God and from His power; and no one was able to enter the temple until the seven plagues of the seven angels were finished." (Revelation 15:5-8) And when the last ruler is slain, the temple of blood shall be gone, "for the Lord God the Almighty and the Lamb are its temple. And the city has no need of the sun or of the moon to shine on it, for the glory of God has illumined it, and its lamp is the Lamb." (Revelation 21:22-23)

∞

Time Shall End (NASB)

To Andrew the Prophet

Completed March 2, 2008

The prophecy said that man would live forever. (Revelation 22:5) And forever means time without end, for its roots are the words "for" and "ever". And "for" is the Germanic word *fore* for "before". And "ever" is the Germanic word *æfre* for "always". Thus forever is "before and always". And God exists before and always, for He "is the Alpha and the Omega, the first and the last, the beginning and the end." (Revelation 22:13) So let man forever "praise the name of the Lord, for He commanded and they were created. He has also established them forever and ever; He has made a decree which will not pass away." (Psalms 148:5-6)

$$\alpha + \omega = \infty$$

Alpha Omega Infinity

And the existence of infinity is beyond the earth's realm. For infinity is derived from *infinitas,* which means "without bounds". But the earth was created with boundaries, and man was created with the bounds of earth's laws. (Deuteronomy 6:8) And even the Son of God was a "bond-servant, and made in the likeness of men." (Philippians 2:7) Yet our heart, soul, and mind can exist without bounds. (Luke 10:28) Thus I will "bind up the testimony, seal the law among my disciples. And I will wait for the Lord who is hiding His face from the house of Jacob; I will even look eagerly for Him." (Isaiah 8:16-17)

And a line exists in just one dimension, yet a line can extend without bounds ($\rightarrow \infty$). And God can extend a line without bounds, for "where were you when I laid the foundation of the earth? Tell Me, if you have understanding, who set its measurements? Since you know. Or who stretched the line on it?" (Job 38:4-5) And a plane exists in two dimensions, yet two lines can extend without bounds. And this universe exists in three dimensions, yet three lines can extend without bounds. And these three lines stretch "negatively" in three dimensions ($-\infty \leftarrow$) . Thus space extends infinitely in six dimensions. "For in six days the Lord made the heavens and the earth, the sea and all that is in them." (Exodus 20:11) And great is our Lord for His dimension has no bounds:

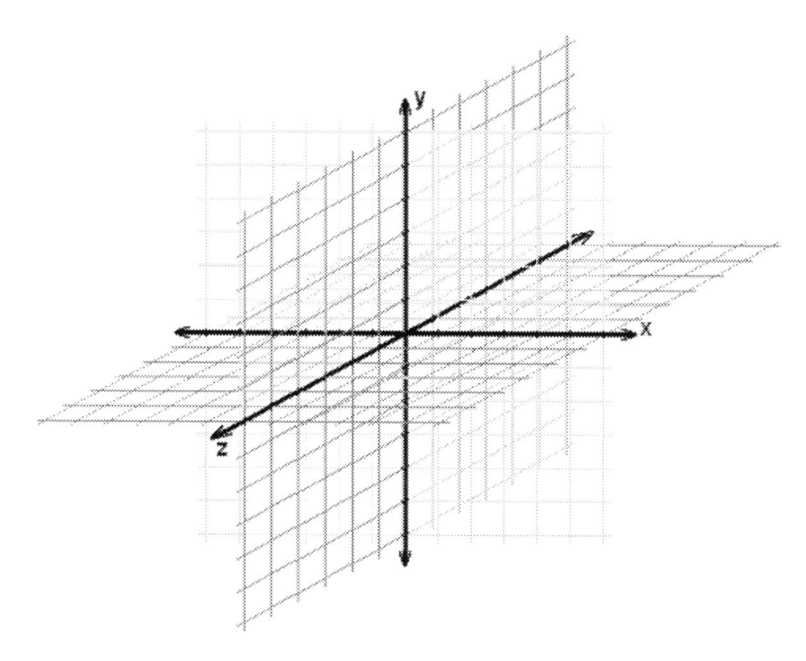

He counts the number of the stars;

He gives names to all of them.

Great is our Lord and abundant in strength;

His understanding is ***infinite***. Psalms 147:4-5

And this universe exists in the dimension of time. Thus infinity has seven dimensions in the realm of this universe. And there are many realms in this universe, for as the Lord said, "in My Father's house are many dwelling places; if it were not so, I would have told you; for I go to prepare a place for you. If I go and prepare a place for you, I will come again and receive you to Myself, that where I am, there you may be also. And you know the way where I am going." (John 14:2-4)

And through the seven days of the week, the greatest hold on mankind has been time. For God created the world in six days, and on the seventh day the whole world would rest. (Genesis 1-2:3) For the Sabbath is a sign of time's hold on mankind. For "it is a sign between Me and the sons of Israel ***forever***; for in six days the Lord made heaven and earth, but on the seventh day He ceased from labor, and was refreshed." (Exodus 31:15-17) But our Lord had forewarned "it is I who put to death and give life. I have wounded and it is I who heal, and there is no one who can deliver from My hand. Indeed, I lift up My hand to heaven, and say, as I live ***forever***." (Deuteronomy 32:39-40) For the seven seals shall be opened, and the seven angels with seven bowls of wrath shall return. And after the seven bowls have been poured, and after the last debt has been paid, then the Lamb shall make war with the ruler of this earth. And at the last battle the eighth shall be slain, for "the beast which was and is not, is himself also an eighth and is one of the seven, and he goes to destruction." (Revelation 17:9-11) For the eighth which is one of the seven is Sabaoth. (Apocrypha of John) For the beast which was and is not is "time". And when the eighth ruler is slain, time shall be no more, and His name shall stand forever more.

$$8 \longrightarrow \infty$$

For His name stands forever and remains undefined. For the Lord's name is YHWH (הוהי). For YHWH is derived from the Hebrew verb *hayah*, which means "I am I". For as the Lord said "I AM WHO I AM... This is My name ***forever***, and this is My memorial-name to all generations." (Exodus 3:13-15) And His Son who was the Word would stand before man, and now "the word of our God stands forever." (Isaiah 40:8) For as was foretold "His name is Jesus. And He will be great and will be called the Son of the Most High; and the Lord God will give Him the throne of His father David; and He will reign over the house of Jacob ***forever***, and His kingdom will have no end." (Luke 1:31-33) And as He said "I am the bread of life. 'Your fathers ate the manna in the wilderness, and they died. This is the bread which comes down out of heaven, so that one may eat of it and not die. I am the living bread that came down out of heaven; if anyone eats of this bread, he will live ***forever***; and the bread also which I will give for the life of the world is My flesh." (John 6:48-51) For the Son in the flesh was nailed to the cross. And on the cross "Pilate also wrote an inscription and put it on the cross. It was written, 'JESUS'. Therefore many of the Jews read this inscription, for the place where Jesus was crucified was near the city; and it was written in Hebrew". (John 19:19-20). For in Hebrew "Jesus the Nazarene King of the Jews" was written ישוץ הנצדי ומדך היהודים (Yeshua' HaNostri U'Melich HaYehudim) or יהוה. Thus the Son's name was written on the cross: יהוה. Thus His name on the cross was written: "I AM WHO I AM". Thus His name stands forever and remains undefined.

Jesus the Nazarene King of the Jews

And the Lord promised all men, that they would inherit His kingdom, if only they would obey His commands. "Oh that they had such a heart in them, that they would fear Me and keep all My commandments always, that it may be well with them and with their sons ***forever***!" (Deuteronomy 5:29) But man in his pride and ignorance, has forsook the Lord's holy name. And now He shall remove the fixed order from this earth. "Thus says the Lord, who gives the sun for light by day and His order of the moon and the stars for light by night, who stirs up the sea so that its waves roar; the Lord of hosts is His name: 'If this fixed order departs from before Me,' declares the Lord, 'Then the offspring of Israel also will cease from being a nation before Me ***forever*** .'" (Jeremiah 31:35-36) For the heavens shall fall and the earth turned to pitch, and man shall be burned into ashes. "For the Lord has a day of vengeance, a year of recompense for the cause of Zion. Its streams will be turned into pitch, and its loose earth into brimstone, and its land will become burning pitch. It will not be quenched night or day; its smoke will go up ***forever***. From generation to generation it will be desolate; none will pass through it ***forever*** and ever." (Isaiah 34:8-10) And "they will not take from you even a stone for a corner nor a stone for foundations, but you will be desolate ***forever***' declares the Lord."(Jeremiah 51:25-26) "For the moth will eat them like a garment, and the grub will eat them like wool. But My righteousness will be ***forever***, and My salvation to all generations."(Isaiah 51:6-8)

Yet the Lord "does not retain His anger ***forever***, because He delights in unchanging love." (Micah 7:18) "For thus says the high and exalted One who lives ***forever***, whose name is Holy, 'I dwell on a high and holy place, and also with the contrite and lowly of spirit in order to revive the spirit of the lowly and to revive the heart of the contrite. For I will not contend ***forever***, nor will I always be angry; for the spirit would grow faint before Me, and the breath of those whom I have made.'" (Isaiah 57:14-16) "'For I am gracious,' declares the Lord; 'I will not be angry ***forever***. Only acknowledge your iniquity, that you have transgressed against the Lord your God and have scattered your favors to the strangers under every green tree, and you have not obeyed My voice,' declares the Lord." (Jeremiah 3:12-13) And "I will betroth you to Me ***forever***; Yes, I will betroth you to Me in righteousness and in justice, in lovingkindness and in compassion, and I will betroth you to Me in faithfulness. Then you will know the Lord." (Hosea 2:18-20) "For behold, I create new heavens and a new earth; and the former things will not be remembered or come to mind. But be glad and rejoice ***forever*** in what I create; for behold, I create Jerusalem for rejoicing and her people for gladness." (Isaiah 65:17-18) "My dwelling place also will be with them; and I will be their God, and they will be My people. And the nations will know that I am the Lord who sanctifies Israel, when My sanctuary is in their midst ***forever***." (Ezekiel 37:25-28) For time shall be forgotten, and God shall dwell in our midst forever. "And there will no longer be any night; and they will not have need of the light of a lamp nor the light of the sun, because the Lord God will illumine them; and they will reign ***forever and ever***." (Revelation 22:5) Amen. All men.

Addendum 1

From ***Signs, Science, and Symbols of the Prophecy***

Christ is the Gate (NASB)

To Andrew the Prophet

Completed August 11, 2007

"I am the way, the truth, and the life. No one comes to the Father except through Me." John 14:6

What does π look like? π is a gate, the design upon which the temples were built. And π is an infinite and transcendental number, a number not understood in finite mortal terms. And Einstein said that if he could comprehend π, then he could comprehend God, for π is the number of God. And it truly is. And what does π look like? It is the narrow gate, for it represents Christ who is the Gate. And do you remember what He said on the Mount? "Enter through the narrow gate; for the gate is wide and the way is broad that leads to destruction, and there are many who enter through it. For the gate is small and the way is narrow that leads to life, and there are few who find it." (Matthew 7:12-13) For we all were created by the Father, and we all have fallen away from the Father. But Christ interceded on our behalf, and gave us the gateway to return to the Father. But to complete the circle back to the Father, we must go through Christ who is the narrow Gate. And what is the formula to complete the circle? The formula is circumference = π • diameter. Circumference is the path to return to the Father, and diameter is the distance you are away from the Father. And Solomon was told when he constructed the temple, that value of π was equal to 3, "ten cubits from one brim to the other; it was completely round... and a line of thirty cubits measured its circumference." (1 Kings 7:23). And obviously that answer was far from complete. For many have come close, but no one on their own has returned to the Father. For this is humanly impossible, for π is an infinite number. Until Christ the Son came down to mankind, for He is the Gate "the way, the truth, and the life. No one can go to the Father except through Me."

(John 14:6) **Christ is π.** When we go through Him, then we return to the Father. But no one can return to the Father except through Him!

Christ = π = The Narrow Gate

Circumference = π • diameter
Circumference = path back to the Father
Diameter = distance between you & the Father
π = 3.14159265358979238472743383279...∞

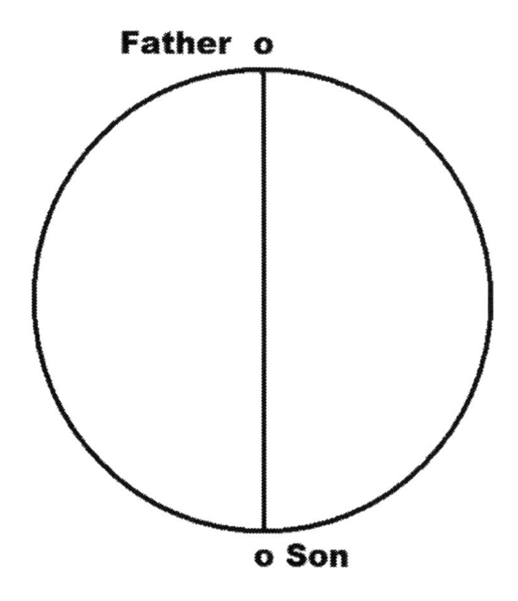

And the circular pattern above is a symbol commonly used in India. For the pattern represents karma, "what goes around, comes around"! All have come from the Father, and all will return to the Father. Amen. All men.

Addendum 2

From ***Signs, Science, and Symbols of the Prophecy***

Entropy, the Law of This Universe, the Law of Satan (NASB)

$$\Delta S = \Delta \text{Satan}$$

To Andrew the Prophet

Completed November 2, 2007

"He will burn up the chaff with unquenchable fire." Matthew 3:12

God's kingdom is ordered

"Truly is not my house so with God? For He has made an everlasting covenant with me, **Ordered** in all things, and secured; For all my salvation and all my desire, Will He not indeed make it grow? But the worthless, every one of them will be thrust away like thorns, Because they cannot be taken in hand" 2 Samuel 23:5-6

"Jesus answered, 'My kingdom is not of this **world**. If My kingdom were of this **world**, then My servants would be fighting so that I would not be handed over to the Jews; but as it is, My kingdom is not of this **realm**.'" John 18:36

"If the **world** hates you, you know that it has hated Me before it hated you. If you were of the **world**, the **world** would love its own; but because you are not of the **world**, but I chose you out of the **world**, because of this the **world** hates you." John 15:18-19

"Command the sons of Israel that they bring to you clear oil from beaten olives for the light, to make a lamp burn continually. Outside the veil of testimony in the tent of meeting, Aaron shall keep it in **order** from evening to morning before the LORD continually; it shall be a perpetual statute throughout your generations. He shall keep the lamps in **order** on the pure gold lampstand before the LORD continually." Leviticus 24:2-4

This universe is ruled by Satan

"Again, the devil took Him to a very high mountain and showed Him all the kingdoms of the **world** and their glory; and he said to Him, 'All these things I will give You, if You fall down and worship me.'" Matthew 4:8-9

"For what will it profit a man if he gains the whole **world** and forfeits his soul? Or what will a man give in exchange for his soul?" Matthew 16:26

"He who loves his life loses it, and he who hates his life in this **world** will keep it to life eternal." John 12:25

"I will not speak much more with you, for the ruler of the **world** is coming, and he has nothing in Me." John 14:30

Entropy and disorder, the laws of Satan, govern this universe

"To the land of darkness and deep shadow, the land of utter gloom as darkness itself, Of deep shadow without **order**, And which shines as the darkness." Job 10:22

"But if you have bitter jealousy and selfish ambition in your heart, do not be arrogant and so lie against the truth. This wisdom is not that which comes down from above, but is **earthly**, natural, demonic. For where jealousy and selfish ambition exist, there is **disorder** and every evil thing." James 3:14-15

Only God allows order in this universe

"Thus says the LORD, Who gives the sun for light by day and the fixed **order** of the moon and the stars for light by night, Who stirs up the sea so that its waves roar; The LORD of hosts is His name" Jeremiah 31:35

We must emulate His order

"But all things must be done properly and in an **orderly** manner." 1 Corinthians 14:40

"They are not of the **world**, even as I am not of the **world**. Sanctify them in the truth; Your word is truth. As You sent Me into the world, I also have sent them into the **world**." John 17:16-18

"Do nothing from selfishness or empty conceit, but with humility of mind regard one another as more important than yourselves; do not merely look out for your own personal interests, but also for the interests of others. Have this attitude in yourselves which was also in Christ Jesus, who, although He existed in the form of God, did not regard equality with God a thing to be grasped, but emptied Himself, taking the form of a bond-servant, and being made in the likeness of men." Philippians 2:3-7

The order of this universe is coming to an end

"'If this fixed **order** departs From before Me,' declares the LORD, 'Then the offspring of Israel also will cease From being a nation before Me forever.'" Jeremiah 31:36

"But now Christ has been raised from the dead, the first fruits of those who are asleep. For since by a man came death, by a man also came the resurrection of the dead. For as in Adam all die, so also in Christ all will be made alive. But each in his own **order**: Christ the first fruits, after that those who are Christ's at His coming, then comes the end, when He hands over the kingdom to the God and Father, when He has abolished all rule and all authority and power. For He must reign until He has put all His enemies under His feet. The last enemy that will be abolished is death." 1 Corinthians 15:20-26

"Then the seventh angel sounded; and there were loud voices in heaven, saying, 'The kingdom of the **world** has become the kingdom of our Lord and of His Christ; and He will reign forever and ever.'" Revelation 11:15

This Universe is Ruled by Disorder or Entropy

The laws of thermodynamics explain the physical laws, the laws of the behavior of energy in this universe. And the first law of thermodynamics is the law of conservation, that energy can be transferred but is neither created nor destroyed.

And energy exists in five different forms: radiant > chemical > physical > electrical > heat. And the highest form is radiant energy, and the lowest form is heat energy, and energy is converted from one form into another.

And the second law of thermodynamics is the law of the degradation of energy, that the quality of energy is degraded irreversibly over time. And when energy is converted from one form to another, degraded energy is lost and is unable to produce further work. And this degraded energy is called **entropy or S.**

The thermodynamic process always proceeds, from a system with a high quality of potential energy and a low amount of entropy, into a system with a low quality of expended energy and a high amount of entropy. And the form of energy with the highest amount of entropy is heat energy, and the transfer of energy can be summarized as follows:

Equation 1:

High quality energy + low entropy → Low quality energy + high entropy

And the mathematical representation of a change in entropy is **ΔS,** where entropy increases as a function of heat, which is mathematically represented by **ΔQ**. Thus, where **T** represents the absolute temperature of the system:

Equation 2:

$$\Delta S = \frac{\Delta Q}{T}$$

Thus entropy increases as heat increases. And in fact, many physicists have hypothesized that the universe is fated to a "heat death", in which all energy ends up as a homogenous distribution of thermal energy, and no further useful energy exists.

For the powers of the universe, follow the laws of entropy. And the powers of this universe are governed, by the powers of its ruler who is Satan. And when the fire of the wrath ensues, entropy or the power of Satan will increase, for as it says, "Woe to the earth and the sea, because the devil has come down to you, having great wrath, knowing that he has only a short time." (Revelation 12:12) And when the radiance of the Son returns, His winnowing fork will gather His wheat, and will throw out the chaff of perdition. "And His winnowing fork is in His hand, and He (high quality energy from the **radiance of the Son**) will thoroughly clear His threshing floor; and He will gather His **wheat** (low entropy) into the barn, but He will **burn** up the **chaff** (high entropy) with **unquenchable fire** (low quality energy)." (Matthew 3:12)

Radiance of the Son + Wheat → Fire + Chaff

From Equation 1:

$$\Delta S = \frac{\Delta Q}{T}$$

From Equation 2:

$$\Delta Satan = \frac{Un\Delta Quenchable}{Fire}$$

And the world in which we live in, is an imperfect image of the kingdom of God. For the laws of this earth are contrary, to the laws that exist in His kingdom. For where there is order in heaven, there is disorder on earth. And where the heavens exceed in all things, the earthly kingdom fails in all things. For King David spoke of the order of God's house, "Truly is not my house so with God? For He has made an everlasting covenant with me, ordered in all things, and secured." (2 Samuel 23:5-6) And the Lord reaffirmed that His house was above, for He said "My kingdom is not of this world. If My kingdom were of this world, then My servants would be fighting so that I would not be handed over to the Jews; but as it is, My kingdom is not of this realm."(John 18:36) And now being servants of Christ, we are not of this world, but are inheritors of His kingdom. And just as Christ died for our sins, so must we suffer for Him, that His heavenly kingdom may be advanced. For as He forewarned us, "if you were of the world, the world would love its own; but because you are not of the world, but I chose you out of the world, because of this the world hates you."(John 15:18-19) But He shall return, for His priests were commanded to keep order in His temple, to "keep the lamps in order on the pure gold lampstand before the LORD continually." (Leviticus 24:2-4) For as the Lord said to His bride, "be dressed in readiness, and keep your lamps lit."(Luke 12:34)

For the realm of this world, is an imperfect image of the kingdom to come, for the ruler of this world is not God, but the ruler of this world is Satan. For when Satan tempted the Lord in the desert, did Satan not tempt Him with the inheritance of this earth? (Matthew 4:8-9) And though Satan would be a fool, to offer this world to a mortal man, the Lord still gave us this solemn warning, "for what will it profit a man if he gains the whole world and forfeits his soul? Or what will a man give in exchange for his soul?" (Matthew 16:26) Thus we are not to place hope in this world, for this world is perishable, but to place our hope in the heavens above, for the heavens are eternal, for "he who loves his life loses it, and he who hates his life in this world will keep it to life eternal." (John 12:25) And before the Lord departed, He promised that Satan would return, "I will not speak much more with you, for the ruler of the world is coming, and he has nothing in Me." (John 14:30) And by our Lord's descent into hell and His rising into heaven, Satan was bound for one thousand years. But Satan was released back into this world, and has ruled it with great power and deception. As it says "when the thousand years were completed; after these things he must be released for a short time." (Revelation 20:3)

And the laws of entropy which govern this world, have a tremendous impact on our view of the universe. For if we truly believe in the laws of entropy, then the only future for mankind is annihilation and death. And when God withheld His blessings from Job, Job succinctly called this world, "the land of darkness and deep shadow, the land of utter gloom as darkness itself, of deep shadow without order, and which shines as the darkness." (Job 10:22) And James spoke of the wisdom which governs this universe, "this wisdom is not that which comes down from above, but is earthly, natural, demonic. For where jealousy and selfish ambition exist, there is disorder and every evil thing." (James 3:14-15) But by God's grace and His loving mercy, we have some semblance of His order, for it is not by the laws of Satan, that the sun gives us light by day, and the moon gives us light by night, but it is by the power of God alone. (Jeremiah 31:35)

And now as servants of God we are taught, to emulate the laws which govern His kingdom, the laws of love and forgiveness and order. For as Paul said, "all things must be done properly and in an orderly manner." (1 Corinthians 14:40) For we are not servants of this world, but were sent in this world as servants of God, to spread the Word of His sacrifice for mankind. (John 17:16-18) And by our servitude, God commands us to be His bond servants, to walk on the earth as Christ did, without selfishness or hatred or conceit, but "regarding one another as more important than ourselves." (Philippians 2:3-7)

But few have followed His commands, and God will remove that which He controls, the order of this world and the universe. For the prophets of old foretold, that the fixed order of this world would end, "'if this fixed order departs from before Me,' declares the LORD, 'then the offspring of Israel also will cease from being a nation before Me forever.'" (Jeremiah 31:36) For as the Lord promised, mankind will never again be destroyed by water, for this time mankind will be destroyed by fire. And the Lord will return and harvest his fruits, "but each in his own order: Christ the first fruits, after that those who are Christ's at His coming, then comes the end, when He hands over the kingdom to the God and Father, when He has abolished all rule and all authority and power. For He must reign until He has put all His enemies under His feet. The last enemy that will be abolished is death." (1 Corinthians 15:20-26) For the author of death is Satan, and the author of life is the Father. For Satan is EVIL, but the Lord reverses all things, so that all men can LIVE. And when the seventh angel has completed his task, Christ the Lamb will be victorious. For no longer will Satan rule this world, for "the kingdom of the world has become the kingdom of our Lord and of His Christ; and He will reign forever and ever." (Revelation 11:15)

Addendum 3

From ***Signs, Science, and Symbols of the Prophecy***

Oil-Currency: Trading for the €nd (NASB)

To Andrew the Prophet

Completed October 16, 2007

"A quart of wheat for a denarius, and three quarts of barley for a denarius; and do not damage the **oil** and the wine." Revelation **6:6**

"And he causes all, the small and the great, and the rich and the poor, and the free men and the slaves, to be given a mark on their right hand or on their forehead, and he provides that no one will be able to buy or to sell, except the one who has the mark, either the name of the beast or the number of his name." Revelation 13:16

Oil-currency is the exchange rate for the end times

The end times were put into motion over a century ago when oil became a major means of influencing the world's currency. For prior to that, the major means of influencing the world's currency was through war. And now the world powers have two means of influencing the world's economy. And what is the world's major currency? It is the US dollar, which accounts for two thirds of all of the world exchange reserves. More than 80% of all foreign exchange transactions and over 50% of all world imports are exchanged in US dollars.

Because the world exchange currency is in US dollars, our government and nation can essentially produce paper currency and receive imports at almost no cost. Our US dollar has little equitable financial backing, except for the fact that it is the currency by which foreign trade is conducted. This can be reflected in the fact that the value of our imports greatly exceeds the value of our exports. In fact just last year, the economic value of our imports was worth more than 50% of our export value.

The major objective of introducing the Euro (€ **for end times**) was to turn the Euro into a reserve currency to compete against the US dollar. And a major means of allowing other countries to compete against the US dollar would be to convert the exchange currency of oil from US dollars to Euros. By forcing oil currency into Euros, the US (the #1 importer of oil) would be forced to exchange US dollars into Euros, and due to the loss of exchange value, this would cause a domino effect on the trade value of the US dollar. Conservative estimates suggest that a conversion to Euros would decrease the dollar value by more than 40%. Ultimately, this would lead to a major crash of the US property and stock markets, thus spiraling the US into a major recession.

And in fact, in November 2000, Saddam Hussein decided to attack the US economy in retaliation for Desert Storm and subsequent trade embargos, by converting Iraq's oil exchange into Euros. This caused a significant decrease in the US dollar exchange rate, and OPEC (Organization of Petroleum Exporting Countries) was threatening to follow suit. Because we could not control a regime like Iraq economically, our government decided to use alternative means to control their oil currency, ...**military force**. Presently, Iran has also converted its oil exchange to the Euro to show support from the "longer horn" (Iran) to the "shorter horn" (Iraq). And historically, this relationship will go down in infamy. "Now the two horns (Iraq & Iran) were long, but one (Mahmoud Ahmadinejad) was longer than the other (Moqtada al-Sadr), with the longer one coming up last" (Daniel 8:3)

†Notation on **ε**: **Epsilon** is the fifth letter of the Greek alphabet. Interestingly, in mathematics, it means a small quantity of anything. i.e. "The cost is only epsilon" And the Euro was introduced into the market because the US was purchasing oil for just an "epsilon" amount. For the cost of oil is "a quart of wheat for a denarius, and three quarts of barley for a denarius." (Revelation 6:6) And **ε** is also utilized in economics as a measurement called elasticity. Elasticity measures how much the demand of a product changes as its price changes. For example, oil economically has a low elasticity, because the economy requires oil regardless of its price. And due

to oil's low elasticity, the oil industry can increase the price of oil or increase their profit margin at will, because the public always has a high demand for its product. And in fact in 2005, the US oil industry had revenues in excess of 1.6 trillion dollars, and over 80% of the revenue was accounted for by the major oil companies or "Big Oil".

Oil-currency is the reason for military and economic force

For those of you who doubt the effect that **oil** has on the world and what seductive powers it has over our leaders, here is a brief history lesson:

Oil is required for almost every aspect of our economic and military power. It is required for transportation (land, sea, and air), heating, agriculture, medications, machinery, and synthetic manufacturing i.e. plastics and rubber. Moreover, oil use has sky rocketed over the past century because of gas-guzzling automobiles, neglect of public transportation, dispersed suburban housing patterns, and excessive consumption of products. At present, the world is consuming 80 million barrels of oil daily. The US is presently the number one consumer and importer of oil at a consumption rate of about 20 million barrels daily. Each US citizen consumes about 1 barrel of oil every 2 weeks. Due to the poverty of the Middle East region, foreign oil has always been much cheaper to produce and transport than domestic oil, thus giving the oil companies a huge profit margin and the incentive to obtain oil overseas rather than domestically. The US government has tried in the past to control oil supplies independently, but due to the opposition of the private sector, it has never been able to do so. Because of that, the government is largely controlled by the interests of the private oil companies, especially the Seven Sisters: Exxon, Mobil, Chevron, Texaco, Gulf Oil, British Petroleum, Shell (British owned). Of principle concern to these companies, is maintaining the security and stability of the Middle East. As we know, this has often entailed using covert and blatantly militaristic schemes to ensure control of certain countries, namely Iraq, Iran, Afghanistan and Kuwait. And now that the US has military control of Iraq, we now have control of 75% of the world oil reserves (i.e. **one-hundred thousand barrels daily** from Caspian reserves via the trans-Afghanistan pipeline and **two-hundred-fifty billion** barrels of Iraqi oil). Clearly, the US hopes to control world oil trade and force opposing nations to purchase oil in US dollars rather than Euros, and thus force nations to hold hundreds of billions of US dollars in reserve for future transactions.

A HISTORY LESSON IN OIL

M = military ε = economics

Highlights involve Middle East

1886	ε	Gas powered automobile invented
1890	ε	US overtakes Great Britain as the major industrial power of the world
1901	ε	Massive oil fields found in Texas
1909	ε	Model T introduced by Ford, drives demand for oil
1914	M	World War I - usage of oil causes massive shortage, thus future military and economic efforts are geared towards securing oil reserves
1920	ε	Oil becomes the fuel of choice for land (automobiles), sea, and air transportation
	ε	Oil becomes critical for agriculture (pesticides and fertilizers)
	ε	US becomes the number one oil exporter and Britain and France rely on the US for oil supplies
	ε	Open door policy for private US oil companies (not the US government) to access foreign oil
	ε	Agreements between British and US oil companies to allocate, fix prices, and monopolize oil
	ε	Mexican dictator Diaz exports oil to the US

	ε	**US & European oil companies begin drilling for Middle Eastern Oil**
1928	ε	**Iraq Petroleum Company established - allows select US oil companies access to Iraqi oil**
1930	ε	Discovery of the great East Texas oil field
	ε	Germany relies on the Soviet Union for oil supplies
	ε	Japan relies on the US for oil supplies
	ε	US switches imports from Mexico to Venezuela due to concern over the Mexican Revolution
	ε	Defeat of the Anglo-American Oil Agreement which would have allowed domestic oil to compete fairly with foreign oil imports
1930	ε	**Standard Oil of California (SOCAL) - now Chevron - controls oil in Bahrain off Saudi Arabia**
1933	ε	**The Texas Company, now Texaco, and SOCAL have oil concession rights in Saudi Arabia**
1935	ε	Dupont invents nylon which is derived from oil products
1936	ε	**The Gulf Oil Company has oil concession rights in Kuwait**
1938	ε	Mexican nationalization, takes over oil production in the Gulf of Mexico, US boycotts Mexican oil
1940	ε	**Roosevelt attempts to set up a government owned oil company to take oil rights in Saudi Arabia. However, private oil companies oppose government control, and the proposal for a government controlled oil company is abandoned.**
	ε	Huge tax breaks for oil companies to import foreign oil
1941	M	Germany attacks Soviet Union to take over Russian oil fields

M Japan attacks US (Pearl Harbor) to assume control of the Netherlands East Indies oil fields (Japan decimates the US Pacific armada which protects this region)

M US supplies oil for the war effort and its European allies, which causes a strain on US oil reserves

1944 M US regains control of the Pacific, Japan's military crippled by oil shortage

M Soviets regain control of oil fields, Germany's military crippled by oil shortage

1946 ε **"Great oil deals" - US oil companies secure control of Middle East oil mediated by Roosevelt**

1947 M Truman Doctrine - US committed to global containment of communism

1948 M Military coup in Venezuela, US continues to work with Venezuelan dictatorship for oil

M **US supports Israeli independence causing alienation from Middle East suppliers; however, the private oil companies distance themselves from the US government and maintain a relationship with the Middle East**

1950 ε Oil is responsible for 29% of the world energy consumption

ε US oil companies dominate the oil-producing regions outside of the Soviet Union

ε US supplies oil to Germany and Japan to spur growth & control recurrence of nationalistic aggression

ε US supplies oil to Western Europe to control Soviet aggression

ε Seven Sisters control 90% of the oil reserves and 75% of the refineries in the world

	ε	Soviet oil production drops drastically and shut out from oil imports from the Middle East
	ε	Oil prices drop due to competition and oversupply of oil
1951	ε	**Iranian nationalization under Massadeq takes over British owned oil companies, Britain calls for an international boycott of Iranian oil, Iran positioned to become a <u>DEMOCRATIC</u> nation**
	ε	Phillips Petroleum develops plastics, derived from oil products
1953	M	**US and Britain organize, finance, and direct a coup against Massadeq and establish a pro-US government under the dictatorship of the Shah of Iran, US oil companies take over control of Iranian oil**
1954	ε	Alternative fuel energy program abandoned by Eisenhower due to pressure from private oil companies
1958	M	**US attacks Lebanon to set up a pro-US regime**
	M	**Brigadier Qassem takes over the British installed king of Iraq, becomes a communist tolerant state**
1956	M	**Suez crisis: Egyptian nationalist Nasser takes over Suez canal (2/3 of oil from the Middle East to Western Europe travels through the canal), US refuses to assist British and French to pressure Egypt into relinquishing control of the Suez, US rather than Europe becomes the controlling foreign power in the Middle East region**
1960	ε	**OPEC (Organization of the Petroleum Exporting Countries) is formed in Baghdad - oil ministers in Iran, Iraq, Kuwait, Saudi Arabia, and Venezuela coalesce to control oil market prices**

	M	**Ayatollah Khomeini expelled from Iran, lives in exile in Iraq under protection of al-Sadr**
1963	M	**Iraq's Qassem overthrown by CIA backed coup, Iraq power assumed by Ba'ath party (Saddam Hussein's party)**
1967	M	**Arab-Israeli Six Day War, US support of Israel causes a strain on US-Arab relations, OPEC raises oil prices 70 percent and embargos US oil shipments, Iraq-US relations severed**
	ε	**Nationalization of Iraqi oil, control is transferred to the Iraq National Oil Company**
1969	M	**Nixon announces the US will rely on and support regional allies, Iran and Saudi Arabia. This results in the US trade of the latest military equipment to the Middle East countries.**
1970	ε	Oil is responsible for 46% of world energy consumption (47% of US energy consumption, 80% of Japan's energy consumption, 64% of Europe's energy consumption)
	ε	US produces only 20% of worlds oil production
	ε	Oil imports rise from 9% in the 1950s to 36% to meet US oil consumption
	ε	**Middle East accounts for more than 40% of the world's oil production**
	ε	**OPEC countries nationalize their oil industries; however, the Seven Sisters receive a huge compensation and maintain their oil access to the Middle East**
1972	M	**Saddam Hussein signs Iraqi-Soviet Friendship Treaty causing US to shift away from Iraqi oil**
	M	**Nixon sells $22 billion in arms sales to the Shah of Iran, US sends military support to Kurds in Northern Iraq**

1973 M **Israel-Palestinian Yom-Kippur war, US alienates Middle Eastern countries due to assistance**

ε **OPEC embargos against the US cause oil shock and severe economic recession**

1974 ε **IEA (International Energy Agency) established - Western nations organize to reduce future reliance on Middle East oil**

1979 M **Iranian revolution, the US friendly Shah is overtaken by Ayatollah Khomeini, establishes the *Islamic fundamentalist government* of present day Iran, results in oil shortage in the US**

ε **Iranian oil shortage causes an increase in inflation, unemployment, and interest rates in the US**

M **Iran-US hostage situation, Iran-Contra negotiations through Reagan and Bush, Sr.**

M **Saddam Hussein assumes control of Iraq, due to loss of control in Iran, US shifts military support to Iraq**

1980 M **Carter doctrine: sets up rapid deployment forces for use of military force in the Middle East. US sells sophisticated military arms and US treasury securities to Saudi Arabia**

M **Iran-Iraq War: US supplies Saddam Hussein in Iraq with military weapons**

M **US gives military assistance to Osama bin Laden to fight against the Soviets in Afghanistan US assistance helps in forming the present day al-Qaeda**

1981 M **Reagan transforms the Rapid Deployment Force into the Central Command**

1984 M **Increased US military support given to Saddam Hussein to oppose Iran**

1986 ε Oil prices collapse due to increased supply and decreased demand

ε Soviet Union suffers greatly due to collapse of oil earnings, eventually leads to collapse of the Iron Wall

1988 M **Saddam Hussein attacks Iranian troops in Northern Iraq, US increases military support**

1990 ε **US imports 25% of oil from the Middle East**

ε Gas prices remain low, energy consumption increases drastically

M **Iraq invades Kuwait to assume control of their oil fields**

1991 M **Operation Desert Storm: 200,000 Iraqi troops killed**

M **Hussein withdraws troops from Kuwait, sets fire to oil wells**

M **Embargos against Iraq following Desert Storm results in deaths of 1 million children due to lack of water, medications, and starvation**

ε Iron Wall collapses due to economic collapse of Soviet Union

1998 ε Oil prices drop to $10 per barrel, causes surge in energy consumption

2000 ε **Sharp increases in oil prices due to rising consumption and OPEC production cuts**

2000 ε **Iraq becomes first OPEC nation to trade oil for Euros**

2001 M **Al-Qaeda under Osama bin Laden attacks the US: NY Trade Towers**

M **Afghanistan war: US establishes new regime in Afghanistan, US builds one-million barrel/day Afghanistan oil pipeline**

2002 M **Massive air strikes on Iraqi targets occurs one month prior to US Congress giving Bush the authority to invade**

M **Bush administration claims Iraq has Weapons of Mass Destruction (WMD)**

M **UN and International Atomic Energy Agency find no evidence of WMD prior to US-Iraq War**

2003 M **Iraq War - also considered by Jihadists to be the final holy war, WW 3**

M **No Weapons of Mass Destruction found**

ε **Lucrative multibillion dollar contracts offered to Halliburton thru Kellogg Brown and Root - a subsidiary of Halliburton ($18.5 billion contract), and Bechtel - multiple connections to White House under Bush Sr.**

2005 M **Iraq approaches civil war**

2007 M 3830 US troops killed; 27,753 US troops injured; 1,087,731 Iraqi civilians killed as of October 2007

ε Cost of war $470 billion as of October 2007

BUSH ADMINISTRATION

And perhaps there is a conflict of interest in the present US White House administration. Here are their credentials:

Dick Cheney, Vice President:
1995-2000: President of Halliburton, Close ties to Shell and Chevron

1992 Halliburton $8 billion contract to put out Kuwait oil fires

2001 Cheney organizes Energy Task Force - advisors consists of executives from ExxonMobil, Chevron, Conoco, Shell Oil, and BP America

2005 Halliburton $1.4 billion billed to US taxpayers unaccounted for

2006 Halliburton surpasses $20 billion in Iraq contracts

Condoleezza Rice, Secretary of State and National Security Advisor:
1991-2000 - Manager of Chevron Oil, International Oil Tanker named after her

Samuel Bodman, Secretary of Energy: named Lee Raymond, former CEO of ExxonMobil, to develop new policy solutions to the US Energy Crisis

Donald Evans, Former Secretary of Commerce: former CEO and chair of Tom Brown, Inc. (a multi-billion dollar oil company)

Steven Griles, Former Secretary of Interior: former executive and lobbyist for Sunoco oil company

Gale Norton, Former Secretary of Interior: former national chairwoman of the Coalition of Republican Environmental Advocates-funded by BP Amoco, former lobbyist for Delta Petroleum - an oil interest group

Thomas White, Former Secretary of the Army: former Vice Chairman of Enron and a large shareholder of company's stock

AND THE DECEIT OF OIL CONTINUES

And the Bush administration made an ill advised attempt to force a new Iraq Oil and Gas Law on February 15, 2007 which was leaked onto the internet and thus states:

Article 5: Management of Petroleum Resources

The **Federal Oil and Gas Council** shall include: "the Chief Executives of important related petroleum companies" which includes the **CEOs of Exxonmobil, Shell, and Chevron Texaco**

"The **Federal Oil and Gas Council** sets the special instructions for negotiations pertaining to **granting rights or signing Development and Production contracts**, and setting qualification criteria for companies."

Article 13: Exploration and Production Contracts

"On the basis of a Field Development Plan prepared and approved in accordance with this Law and the relevant contract, INOC and other holders of an Exploration and Production right may retain the exclusive right to develop and produce Petroleum within the limits of a Development and Production Area for a period to be determined by the **Federal Oil and Gas Council** varying from fifteen (15) to twenty (20) years, not exceeding twenty (20) years dating from the date of approval of the Field Development Plan, depending on considerations related to optimal oil recovery and utilization of existing infrastructure. In cases which for technical and economic considerations warrant longer Production period, the **Federal Oil and Gas Council**, on newly negotiated terms, has the authority to grant an extension not exceeding five (5) years."

"The appointment of an Operator shall be approved by the Designated Authority, and the procedures for such appointment are stated in the initial Contract, and according to the

guidelines issued by the **Federal Oil and Gas Council**, and the Operator should be named in the initial Contract."

Article 40: Existing Contracts

"The **Designated Authority** in the Kurdistan Region will take responsibility to review all

existing Exploration and Production contracts with any entity before this law enters into force to ensure harmony with the objectives and general provisions of this law to obtain maximum economic returns to the people of Iraq, taking into consideration the prevailing circumstances at the time at which those contracts were agreed, and in a period not exceeding three (3) months from the date of entry into force of this law. The **Panel of Independent Advisors** will take responsibility to assess the contracts referred to in this Article, and their opinion shall be binding in relation to these contracts."

The **"Designated Authority"** and **"Panel of Independent Advisors"** is not disclosed.

So, we will let the Bush administration and Big Oil claim the oil and do as the Lord commands ... "save the OIL and the WINE", the fornication of oil and the blood of mankind, for as He promised "Truly I say to you, you will not come out of there until you have paid up the last cent." (Matthew 5:26)

LORD, MAY THEIR OIL BURN IN THE FIRE.

Addendum 4

From ***Signs, Science, and Symbols of the Prophecy***

Iota: The Number of the Holy Spirit

To Andrew the Prophet

Completed January 22, 2008

ι Iota, The Holy Spirit (Numerical value = 10)

"But the Helper, the **Holy Spirit**, whom the Father will send in My name, He will teach you all things, and bring to your remembrance all things that I said to you." (John 14:26)

Iota is a beautiful number, because like the Holy Spirit, it may appear insignificant and is often invisible, but in actuality is ubiquitous and powerful. "Where can I go from Your Spirit? Or where can I flee from Your presence? If I ascend to heaven, You are there; if I make my bed in Sheol , behold, You are there. If I take the wings of the dawn, if I dwell in the remotest part of the sea, even there Your hand will lead me, and Your right hand will lay hold of me. If I say, 'Surely the darkness will overwhelm me, and the light around me will be night,' even the darkness is not dark to You, and the night is as bright as the day. Darkness and light are alike to You." (Psalms 139:7-12)

Iota may be the smallest letter of the Greek alphabet, but in Greek numerals iota represents the powerful value of 10. The number ten represents perfection and productivity. In like manner, the Holy Spirit is God's perfection, and gives His servants great productivity. In mathematics, it is the root of our numerical system. The reason for this choice is assumed to be that humans have ten fingers or digits. The number 10 is a powerful and productive number because the value of any integer can be increased tenfold simply by adding a zero to the end of it. And interestingly, the Chinese numeral for ten is 十 which resembles a cross. And of course, the number 10 is repeated throughout the history of God. For the Israelites were instructed to give one-tenth of their produce to God, and God's finger wrote on stone the Ten Commandments, and God inflicted Egypt with Ten Plagues. Therefore, just as the iota represents numerical power, perfection and productivity, the Holy Spirit represents Godly power and perfection and spiritual productivity. In other words, the gifts of the Spirit produce great power through the fruits of our labor. "For our gospel did not come to you in word only, but also in power and in the Holy Spirit." (1 Thessalonians 1:5)

Iota is also used as a marker of pronunciation by diphthongs. Before the development of the classical Greek alphabet, there was no way of distinguishing between long and short vowels. Thus diphthongs, or iotas, were added to the words to distinguish the pronunciation of these vowels. And these iotas can now be seen throughout various languages including the Czech, Dutch, English, Faroese, Finnish, French, German, Hungarian, Icelandic, Italian, Latvian, Northern Sami, Norwegian, Portuguese, Romanian, and Spanish languages. So just as the iota has ubiquitously influenced the pronunciation of many languages, so has the Holy Spirit ubiquitously penetrated the voice of all the nations. "But whatever is given you in that hour, speak that; for it is not you who speak, but the Holy Spirit." (Mark 13:11) "And they were all filled with the Holy Spirit and began to speak with other tongues, as the Spirit gave them utterance." (Acts 2:4)

The iota has also penetrated the written languages and the artistry of the world. The iota was written as an 'iota subscript' in the original languages. The 'iota subscript' is a way of writing the letter iota as a small vertical stroke beneath a vowel. Eventually, they became incorporated as serifs, or non-structural details, on the ends of strokes that make up letters and symbols. These embellishments were also incorporated into the Asian and Aramaic written languages. Thus, just as the iota has penetrated all the written and artistic languages, so has the Holy Spirit penetrated the written and artistic languages of all the nations. "And I have filled him with the Spirit of God, in wisdom, in understanding, in knowledge, and in all manner of workmanship, to design artistic works, to work in gold, in silver, in bronze, in cutting jewels for setting, in carving wood, and to work in all manner of workmanship." (Exodus 31:3-5) And like the letter of the languages of the nations, our hearts have become the letter of the Spirit. "You are our letter, written in our hearts, known and read by all men; being manifested that you are a letter of Christ, cared for by us, written not with ink but with the Spirit of the living God, not on tablets of stone but on tablets of human hearts." (2 Corinthians 3:2-3)

"For assuredly, I say to you, till heaven and earth pass away, one jot or one tittle will by no means pass from the law till all is fulfilled." (Matthew 5:18) And this is the one scriptural passage that directly refers to the letter iota. "Jot" translates to "iota" in Greek, and is paraphrased as "not one iota of difference". This was the foundation of the theological debate at the First Council of Nicaea regarding the nature of the Trinity. The argument centered on which of two Greek words, differing only by a single "iota", should be used to distinguish Jesus' relationship in the Holy Trinity. One word "homoousious" meant that Jesus was of the same substance as God the Father, and the other "homo***i***ousious" meant that Jesus was of a similar substance. The verdict, of course, was that they are "homoousious". **They are all of the same substance ...the Father, the Son, and the Holy Spirit are Three in One: the Holy Trinity.**

Finally in mathematics, the imaginary unit, iota, allows the real numerical system to be extended into the complex numerical system. Real numbers are those numbers that can be represented by "real" numerical values; in other words, numbers that exist in the "real world". Complex numbers are also called imaginary numbers, and represent those numbers that were once thought to be nonexistent, but presently are utilized to understand quantum physics and the theory of relativity. The numerical value of an imaginary unit is:

$$\iota = \sqrt{-1}$$

which is an impossible value by real number mathematics.

In other words, they are numbers that presently do not exist in our physical world, but do exist in another realm (a parallel universe); a realm that we do not yet understand. Thus, as the imaginary unit, iota, in mathematics is used to transcend numbers from a real set of values into an imaginary set of values, so does the Holy Spirit transcend mankind from the mortal realm of this earth, to the eternal realm of the Father's kingdom. "Immediately I was in the Spirit; and behold, a throne was standing in heaven, and One sitting on the throne." (Revelations 2:4)

Addendum 5

Nostradamus' Quatrain of George W Bush, Jr.

To Andrew the Prophet

Completed January 24, 2008

Les Propheties

(Century 10 Quatrain 72-73)

Michel de Nostredame

The year 1999, seventh month,
From the sky will come a great King of Terror:
To bring back to life the great King of the Mongols,
Before and after Mars to reign by good luck.

The present time together with the past
Will be judged by the great Joker:
The world too late will be tired of him,
And through the clergy oath-taker disloyal.

The year 1999, seventh month

Mickey Herskowitz, author and journalist, was given unrestrained access to then presidential candidate George W. Bush Jr.. According to Herskowitz, Bush was already talking privately about the political benefits of attacking Iraq two years before September 11th occurred. Bush's advisors felt Herskowitz was too candid in his disclosure of information about George W. Bush Jr. and fired him in July, the ***seventh month of 1999***.

And as quoted by Herskowitz regarding his accounting of George W Bush Jr., "his lawyer called me and said, 'Delete it. Shred it. Just do it.'" For "He was thinking about invading Iraq in 1999," reported Herskowitz. "It was on his mind. He said to me: 'One of the keys to being seen as a great leader is to be seen as a commander-in-chief.'

And he said, 'My father had all this political capital built up when he drove the Iraqis out of Kuwait and he wasted it.' He said, 'If I have a chance to invade, if I had that much capital, I'm not going to waste it. I'm going to get everything passed that I want to get passed and I'm going to have a successful presidency.'"

And according to Herskowitz, George W. Bush's position is deeply rooted in the opinions of now vice-president Dick Cheney. For Dick Cheney was quoted as advising George W Bush Jr., "Start a small war. Pick a country where there is justification you can jump on, go ahead and invade."

To bring back to life the great King of the Mongols,
Before and after Mars to reign by good luck.

Khan Conquests Middle East (1222-1227)

The King of the Mongols, Genghis Khan was infamous for his ruthless tactics in overtaking Asia and the Middle East. In fact, legend purports that he was born with a blood clot clinched in his fist.

Khan had three brothers and two sisters. Because his father was the chieftain, Khan inherited the role as the leader of the Mongols. In his vision, he saw the Mongols as the greatest people on earth, and felt that he was ordained by god. After conquering southern China and making alliances with northern China, Khan extended his territory by attacking west through Afghanistan into Iraq. He intelligently did not mutilate his enemies, but preferred to attack and kill them from a distance. After his conquests, he sought to increase trade, especially in oils and other commodities, through China and the Caspian Sea and through Arabia and India.

George Bush Jr. had three brothers and two sisters. Because his father was the US president, George Bush Jr. inherited the role as the leader of the US. In his vision, he saw America as the greatest people on earth, and felt that he was ordained by god. Bush extended his territory by attacking west through Afghanistan into Iraq. After his conquests, he sought to increase trade, especially in oil through the Caspian Sea and through Saudi Arabia.

And through the clergy oath-taker disloyal

In the televised Republican presidential debate in Iowa in December of 1999, all of the participating candidates were asked "What political philosopher or thinker do you most identify with and why?" And George Bush Jr.'s response was "Christ, because He changed my heart". And He claimed in his book *Charge to Keep: My Journey to the White House,* that through the guidance of Billy Graham he came to know the Lord. But when Billy Graham was asked about this spiritual conversion by NBC's Brian Williams, he declined to corroborate Bush's account. "I've heard others say that (I converted Bush), and people have written it, but I cannot say that," Graham said. "I was with him and I used to teach the Bible at

Kennebunkport to the Bush family when he was a younger man, but I never feel that I in any way turned his life around."

Well maybe Billy Graham didn't, but on that Day the Lord surely will.

Addendum 6

The Sixth Seal by Nostradamus

To Andrew the Prophet

Completed February 5, 2008

Les Propheties
(Century 1 Quatrain 27)

Michel de Nostredame

Earth-shaking fire from the center of the earth.
Will cause the towers around the New City to shake,
Two great rocks for a long time will make war,
And then Arethusa will color a new river red.
(And then areth USA will color a new river red.)

Earth-shaking fire from the center of the earth.
Will cause the towers around the New City to shake,
Two great rocks for a long time will make war

There is recent scientific evidence from drill core sampling in Manhattan, that the southern peninsula is overlapped by several tectonic plates. Drill core sampling has been taken from regions south of Canal Street including the Trade Towers' site. Of particular concern is that similar core samples have been found across the East River in Brooklyn. There are also multiple fault lines along Manhattan correlating with north-northwest and northwest trending neo-tectonic activity. And as recently as January and October of 2001, New York City has sustained earthquakes along these plates.

For there are "two great rocks" or tectonic plates that shear across Manhattan in a northwestern pattern. And these plates "for a long time will make war", for they have been shearing against one other for millions of years. And on January 3 of 2010, when they make war with each other one last time, the sixth seal shall be opened, and all will know that the end is near.

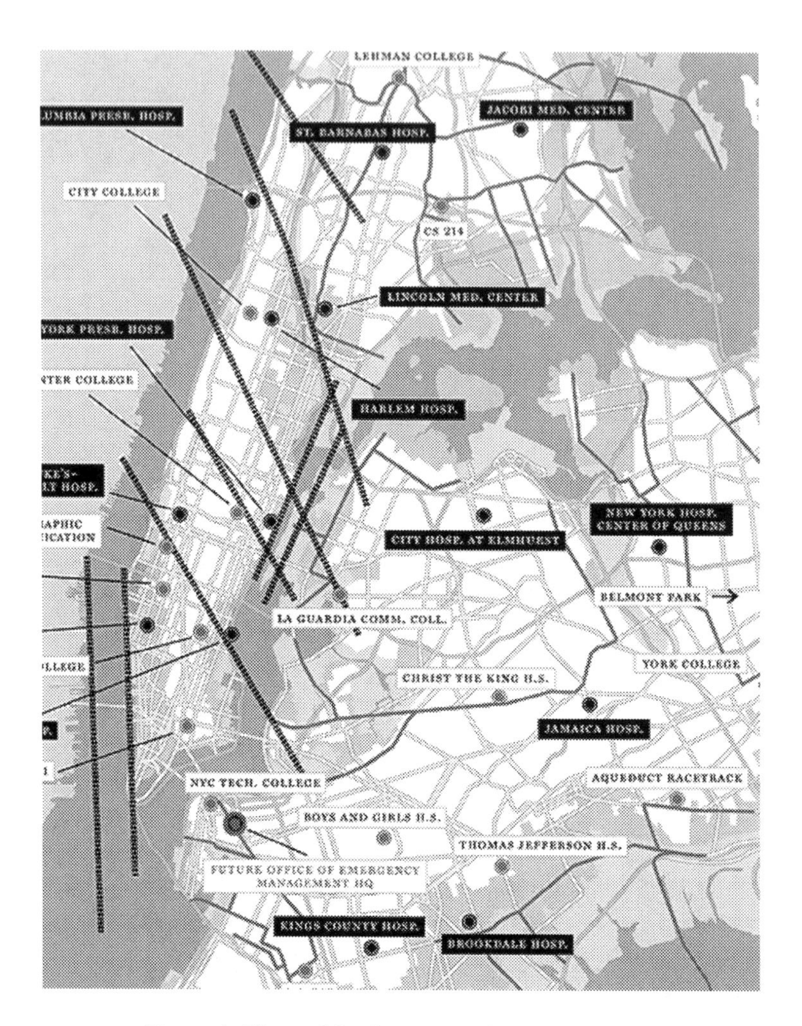

Tectonic Plates Manhattan and Brooklyn, NY

And then Arethusa will color a new river red.

Arethusa is a Greek mythological figure, a beautiful huntress and a follower of the goddess Artemis. And like Artemis, Arethusa would have nothing to do with me; rather she loved to run and hunt in the forest. But one day after an exhausting hunt, she came to a clear crystal stream and went in it to take a swim. She felt something from beneath her, and frightened she scampered out of the water. A voice came from the water, "Why are you leaving fair maiden?" She ran into the forest to escape, for the voice was from Alpheus, the god of the river. For he had fallen in love with her and became a human to give chase after her. Arethusa in exhaustion called out to Artemis for help, and the goddess hid her by changing her into a spring. But not into an ordinary spring, but an underground channel that traveled under the ocean from Greece to Sicily. But Alpheus being the god of the river, converted back into water and plunged down the same channel after Arethusa. And thus Arethusa was captured by Artemis, and their waters would mingle together forever.

And of great concern is that core samples found in train tunnels beneath the Hudson River are identical to those taken from southern Manhattan. Furthermore, several fault lines from the 2001 earthquakes were discovered in the Queen's Tunnel Complex, NYC Water Tunnel #3. And a few years ago, a map of Manhattan drawn up in 1874 was discovered, showing a maze of underground waterways and lakes. For Manhattan was once a marshland and labyrinth of underground streams. Thus when the sixth seal is broken, the subways of the New City shall be flooded be Arethusa: the waters from the underground streams and the waters from the sea. And Arethusa shall be broken into two. And then Arethusa will color a new river red.

***Waterways of NYC* 1874**

And then areth USA will color a new river red.

For Arethusa broken into two is ***areth USA***. For areth (αρετη) is the Greek word for values. But the values of the USA are not based on morality, but on materialism and on wealth. Thus when the sixth seal is opened, Wall Street and our economy shall crash and "areth USA", the values of our economy shall fall "into the red." "Then the kings of the earth and the great men and the commanders and the rich and the strong and every slave and free man hid themselves in the caves and among the rocks of the mountains; and they said to the mountains and to the rocks, 'Fall on us and hide us from the presence of Him who sits on the throne, and from the wrath of the Lamb; for the great day of their wrath has come, and who is able to stand?'" (Revelation 6:15-17)

Addendum 7

Sadr City and Moqtada al-Sadr

Nostradamus and the Prophet Daniel

To Andrew the Prophet

February 16, 2008

Les Propheties

(Century 1 Quatrain 27)

Michel de Nostredame

At forty-five degrees, the sky will burn,
Fire approaches the great new city,
Immediately a huge scattered flame leaps up
When they want to have verification from the Normans.

At forty-five degrees, the sky will burn,
Fire approaches the great new city

After the fall of Baghdad to the coalition forces, Moqtada al-Sadr emerged as a militant voice against the occupation. His stronghold is in Sadr City, named after his father Imam Sadiq al-Sadr who was executed in 1999. The al-Sadr name is highly revered because of their holy lineage, nationalistic ideals, and the martyrdom of their bloodline. Moqtada al-Sadr's two brothers and his uncle Imam Baqir al-Sadr were also assassinated with assistance from the US administration. His army, the Jaish Al-Mahdi or Mahdi army, consists of about two hundred-thousand military loyalists to al-Sadr. The name of this resistance force refers to the Mahdi, an imam who is prophesied by Shi'a Muslims to appear in messianic form during the last days of the world.

Baghdad, Iraq   March 2003

And Sadr City is a suburb of east of Baghdad which is located at ***forty-five degrees*** east latitude. And through the deception of the Bush administration, the sky would burn at forty-five degrees, in preparation before the fall of the New City.

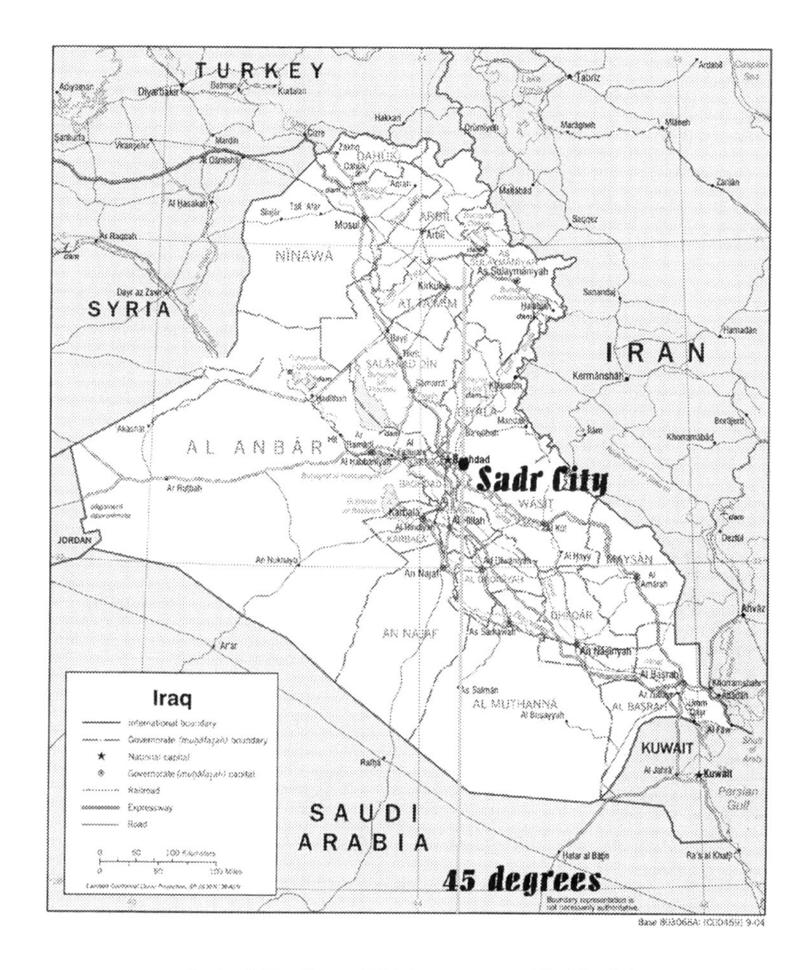

Sadr City, Iraq (45 degrees east latitude)

Immediately a huge scattered flame leaps up
When they want to have verification from the Normans

In 1906, the Norman Crusades invaded the Holy Land, and set up a Norman principality in ***Antioch*** near Jerusalem. For the accounting that Nostradamus was referring to was the prophecy according to Daniel:

"Then the male goat magnified himself exceedingly. But as soon as he was mighty, the large horn was broken; and in its place there came up four conspicuous horns toward the four winds of heaven. Out of one of them came forth a rather small horn which grew exceedingly

great toward the south, toward the east, and toward the Beautiful Land. It grew up to the host of heaven and caused some of the host and some of the stars to fall to the earth, and it trampled them down. It even magnified itself to be equal with the Commander of the host; and it removed the regular sacrifice from Him, and the place of His sanctuary was thrown down. And on account of transgression the host will be given over to the horn along with the regular sacrifice; and it will fling truth to the ground and perform its will and prosper. Then I heard a holy one speaking, and another holy one said to that particular one who was speaking, 'How long will the vision about the regular sacrifice apply, while the transgression causes horror, so as to allow both the holy place and the host to be trampled?' He said to me, 'For 2,300 evenings and mornings; then the holy place will be properly restored.'" Daniel 8:8-14

"In the latter period of their rule, when the transgressors have run their course, a king will arise, insolent and skilled in intrigue. His power will be mighty, but not by his own power, and he will destroy to an extraordinary degree and prosper and perform his will; he will destroy mighty men and the holy people. And through his shrewdness he will cause deceit to succeed by his influence; and he will magnify himself in his heart, and he will destroy many while they are at ease. He will even oppose the Prince of princes." Daniel 8:23-25

For the first accounting of this prophecy was before the coming of Christ the Lamb. For "the male goat that magnified himself exceedingly" was Alexander the Great. And the "four conspicuous horns toward the four winds of heaven" were the four the empires that would rule over the Holy Land. For the four empires were the Macedonian Empire, the Seleucids of Syria, the Ptolemies of Egypt, and the Roman Empire. And ***Antiochus Epiphanes,*** a successor to Alexander, was the "small horn" that grew out of the second empire, the Seleucids of Syria. Antiochus grew "toward the south, toward the east, and toward the Beautiful Land". He "magnified itself to be equal with the Commander of the host; and it removed the regular sacrifice from Him, and the place of His sanctuary was thrown down." For Antiochus attempted to impose his Hellenistic cults on Israel, and in 168 BC even dared to occupy Jerusalem. He entered the Holy of Holies and desecrated the Temple with unclean sacrifices. He erected a statue of Jupiter in the sanctuary, and plundered the treasures of the Temple. But following the denigration of the Temple, the Maccebean revolt occurred, and Israel regained their independence, and the holy place was properly restored.

And the fourth king that arrived at the end of the Roman Empire was Herod the Great. And Herod would destroy to an extraordinary degree, and would even murder the children of the holy people. "When Herod saw that he had been tricked by the magi, he became very enraged, and sent and slew all the male children who were in Bethlehem and all its vicinity, from two years old and under, according to the time which he had determined from the magi." (Matthew 2:16) For he would even oppose the Prince of princes, but "an angel of the Lord appeared to Joseph in a dream and said, 'Get up! Take the Child and His mother and flee to Egypt, and remain there until I tell you; for Herod is going to search for the Child to destroy Him.'" (Matthew 2:13)

And the second accounting of the prophecy of Daniel shall be repeated before the Lion returns. For the four kingdoms that would rule following our Lord's sacrifice, were the Caliphate Empire, the Holy Crusades, the rulers of Egypt, and the Ottoman Empire. And the "small horn" of the Normans arose from the empire of the Holy Crusades. And the Normans grew "toward the south, toward the east, and toward the Beautiful Land" and they would set up their empire in <u>***Antioch***</u>. And they would even throw down "the place of His sanctuary", and destroy the city of Jerusalem. But the Normans were soon cast from the Holy Land. And following the Ottoman Empire, Israel would be restored as an independent state. But "the holy place and host" continue to be trampled to this day. For Jerusalem no longer holds the true temple of God, but is now the outer court of the temple. "Leave out the court which is outside the temple and do not measure it, for it has been given to the nations; and they will tread under foot the holy city for forty-two months." (Revelation 11:2-3) But "a king will arise, insolent and skilled in intrigue", for "a huge scattered flame leaps up", for the beast of the earth Moqtada al-Sadr, the Great Herod of old has returned.

Addendum 8

The Second Woe To Andrew the Prophet

Completed February 17, 2008

Les Propheties

(Century 9, Quatrain 83)

Michel de Nostredame

The Sun in 20 degrees Taurus
There will be a great earthquake;
the great theater full up will be ruined.
Darkness and trouble in the air, on the sky and land,
When the infidel calls upon God and the Saints.

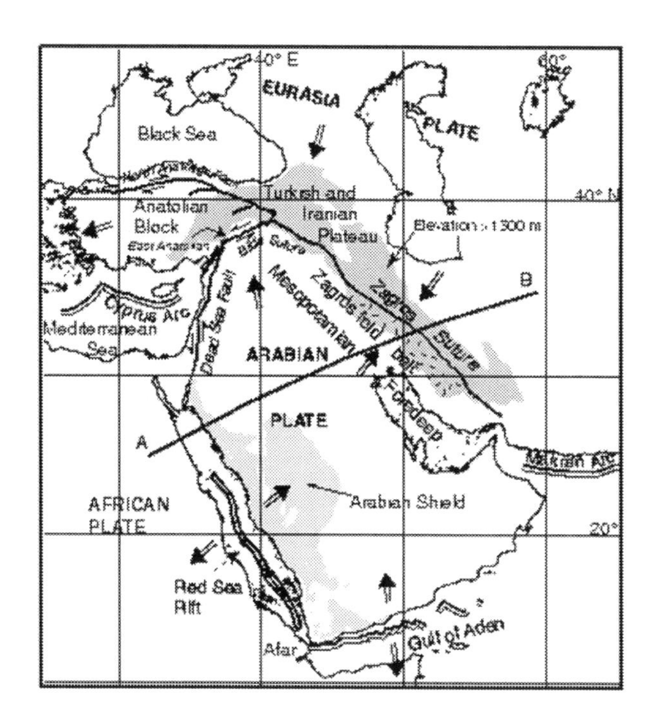

The Second Woe

The Sun in 20 degrees Taurus
There will be a great earthquake; the great theater full up will be ruined.

"Then they went up into heaven in the cloud, and their enemies watched them. And in that hour there was a great earthquake, and a tenth of the city fell; seven thousand people were killed in the earthquake, and the rest were terrified and gave glory to the God of heaven. The second woe is past; behold, the third woe is coming quickly." (Revelation 11:12-14)

"And then the sign of the Son of Man will appear in the sky, and then all the tribes of the earth will mourn, and they will see the Son of Man coming on the clouds of the sky and great glory." (Matthew 24:30)

And on that fateful day the Son of Man will appear, and there will be a great earthquake in Jerusalem, and the great theater of mankind will be ruined. For the Dead Sea Rift is a fault line, that extends through the Dead Sea and under Jerusalem, and into Turkey into the ***Taurus*** Mountains.

Darkness and trouble in the air, on the sky and land,
When the infidel calls upon God and the Saints.

"One day later, the Beast from the earth will miraculously
Emerge from Mount Safaa in Makkah, causing a splint in the ground"

(Qiyamaah, the prophecy of Islam)

And the following day the beast from the earth (Moqtada al-Sadr), shall emerge from the Hajj in Mecca. And from Mount Safaa in Mecca, darkness will cover the sky and the land. For the second woe shall include a quake along the Arabian Plate. For the infidel will call upon God and the Saints, but God will not listen to the son of perdition. And Arabia and the Sea shall be split into two, from Taurus to Mecca at ***20 degrees*****.**

Addendum 9

The Great Enemy To Andrew the Prophet

Completed January 17, 2008

You know me well but soon won't be,
For men have sought to capture me.
But when they seek my hold too long,
By their pursuit they make me strong.
For in my name the ancients lie,
And in my strength they all must die.

The paradox is I'm surreal,
For am I proper or am I real?
I am myself and not defined.
Men can't see me for they are blind.
And all the laws are in my hand.
The end draws near so understand,
The heat will soar and mass will fall,
And I will crawl at my last call.

Yet few men know the truth in me,
For though they think that I am free,
And hope to have me in their hold,
It is their life that I control.

The brightest minds cannot agree,
Which laws define my property.
For Newton though not resolute,
Has called my substance absolute.
And Einstein laws hold like a sieve,
And still he calls me relative.

But some do know the truth in me,
For through the Lamb God sets them free.
And as the Shepherd calls His fold,
The Son has freed them from my hold.

For He's the One who shall slay me,

For in His strength He sets men free.

Though I seem strong, through Him I'm weak.

Though I seem long, my life is bleak.

And never ever say forever,

For Satan's tricks and lies are clever.

But all the foul that they

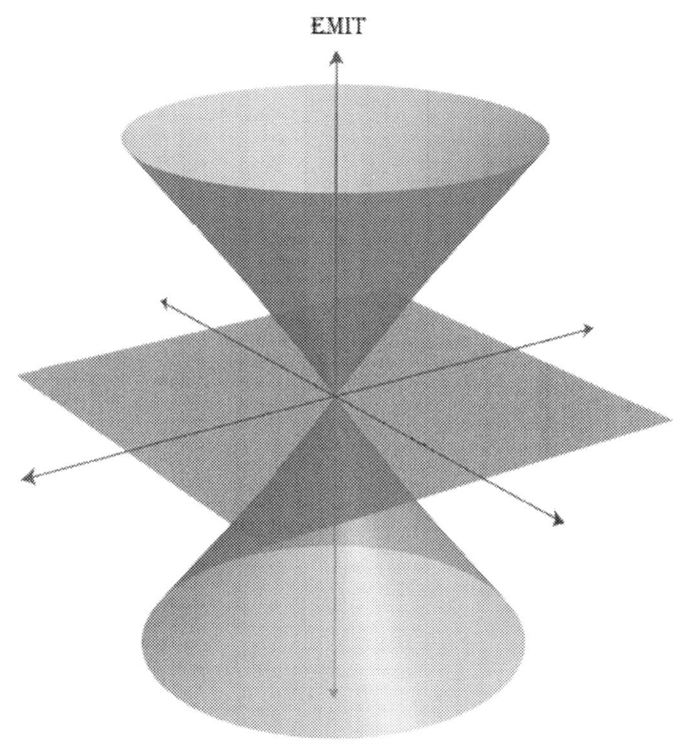